Software Theory

Media Philosophy

Series Editors: Eleni Ikoniadou, Lecturer in Media and Cultural Studies at the London Graduate School and the School of Performance and Screen Studies, Kingston University and Scott Wilson, Professor of Cultural Theory at the London Graduate School and the School of Performance and Screen Studies, Kingston University

The Media Philosophy series seeks to transform thinking about media by inciting a turn towards accounting for their autonomy and 'eventness', for machine agency, and for the new modalities of thought and experience that they enable. The series showcases the 'transcontinental' work of established and emerging thinkers whose work engages with questions about the reshuffling of subjectivity, of temporality, of perceptions and of relations vis-à-vis computation, automation, and digitalisation as the current 21st century conditions of life and thought. The books in this series understand media as a vehicle for transformation, as affective, unpredictable, and non-linear, and move past its consistent misconception as pure matter-of-fact actuality.

For Media Philosophy, it is not simply a question of bringing philosophy to bear on an area usually considered an object of sociological or historical concern, but of looking at how developments in media and technology pose profound questions for philosophy and conceptions of knowledge, being, intelligence, information, the body, aesthetics, war, death. At the same time, media and philosophy are not viewed as reducible to each other's internal concerns and constraints and thus it is never merely a matter of formulating a philosophy of the media; rather the series creates a space for the reciprocal contagion of ideas between the disciplines and the generation of new mutations from their transversals. With their affects cutting across creative processes, ethico-aesthetic experimentations and biotechnological assemblages, the unfolding media events of our age provide different points of intervention for thought, necessarily embedded as ever in the medium of its technical support, to continually re-invent itself and the world.

> *"The new automatism is worthless in itself if it is not put to the service of a powerful, obscure, condensed will to art, aspiring to deploy itself through involuntary movements which none the less do not restrict it".*

Eleni Ikoniadou and Scott Wilson

Titles in the Series

Software Theory, by Federica Frabetti
Media After Kittler, by Eleni Ikoniadou and Scott Wilson
Cypherpunk Philosophy, by Paul J. Ennis

Software Theory

A Cultural and Philosophical Study

Federica Frabetti

London • New York

Published by Rowman & Littlefield International, Ltd.
Unit A, Whitacre Mews, 26-34 Stannary Street, London SE11 4AB
www.rowmaninternational.com

Rowman & Littlefield International, Ltd. is an affiliate of Rowman & Littlefield
4501 Forbes Boulevard, Suite 200, Lanham, Maryland 20706, USA
With additional offices in Boulder, New York, Toronto (Canada), and London
(UK)
www.rowman.com

British Library Cataloguing in Publication Information Available
A catalogue record for this book is available from the British Library

ISBN: HB 978-1-78348-196-5
ISBN: PB 978-1-78348-197-2
ISBN: EB 978-1-78348-198-9

Library of Congress Cataloging-in-Publication Data

Frabetti, Federica, 1966– author.
Software theory : a cultural and philosophical study / Federica Frabetti.
pages cm. — (Media philosophy)
Includes bibliographical references and index.
ISBN 978-1-78348-196-5 (cloth : alk. paper)—ISBN 978-1-78348-197-2 (pbk. : alk.
paper)—ISBN 978-1-78348-198-9 (electronic)
1. Software engineering—Social aspects. 2. Software architecture—Social aspects. I. Title.
QA76.758.F725 2014
005.1—dc23
2014033084

Printed in the United States of America

Contents

Acknowledgements

This book has been inspired by almost fifteen years of 'writing' software as a software engineer for telecommunications. The people I wrote software with are too many to be named; nevertheless, I would like to thank them here because I learnt so much from them while also having so much fun.

The origins of this project lie with my time as a doctoral student at Goldsmiths, University of London, which provided an exciting and stimulating environment for the much more critical questioning of technology. First and foremost I am grateful to Joanna Zylinska for being an exceptional guide and a demanding interlocutor while I was at Goldsmiths, and for remaining a mentor and a friend long afterwards. Her work remains for me a model of intellectual and political engagement. I would also like to express my gratitude to all those who have read various parts of this book and immensely enriched it with their comments and critiques: David Boothroyd, Scott Dexter, Gary Hall, Janet Harbord, Sarah Kember, David Morley, and the late and much missed Mark Poster. I am especially grateful to Gary Hall for proposing *Software Theory* as the title of this book and for sharing with me some of his tremendous insights on technology (not least the one about 'quasi-functioning' software).

Many years ago I had the privilege to discuss the early stages of my project with N. Katherine Hayles at a Masterclass organized by Rosi Braidotti at the University of Utrecht. Although over time my thought on technology has taken a different direction from Hayles's, I maintain the greatest admiration for her work and feel fortunate to have remained in dialogue with her throughout the years. I am grateful to my colleagues and students at Oxford Brookes University for providing a lively environment for the discussion of my thoughts on philosophy and technology and for supporting the realization of this book through various research grants. I am also indebted to a number of other universities which have lent hospitality to my ideas, including Columbia University, New York; Brooklyn College, CUNY; Coventry University, and the University of Naples 'L'Orientale' (Italy). I am very grateful to the editors and reviewers at Rowman & Littlefield International for all their help and support with this project.

Parts of this book have been published as articles. An earlier version of Chapter 1, entitled 'Rethinking the Digital Humanities in the Context of Originary Technicity', acted as an opening to the special issue of *Culture Machine, The Digital Humanities beyond Computing*, vol. 12 (2011), which I edited. An early version of two sections of Chapter 3 and 4 appeared as part of another article, entitled 'Does It Work? The Unforeseeable Consequences of Quasi-Failing Technology', *Culture Machine* (*Creative Media* 11 [2009], http://www.culturemachine.net).

I would like to express my gratitude to family and friends for their support through the years. Thanks to Roberto Camano for providing early help with grammars and compilers. Very special personal thanks go to Liana Borghi and Tiziana Iannucci, for changing my life in so many ways and for encouraging me during the early stage of my academic career.

Finally, all this would not have been possible without the unconditional love and support of Carla Marzocchi, my mother. This book is dedicated to her.

Introduction

This book is inspired by my experience of more than a decade of designing software for telecommunications for a number of companies across two continents. I was fortunate enough to have witnessed, and made a small contribution to, the birth of the second generation of mobile telephony, or GSM (a geological stratum of current G4/IMT Advanced networks).[1] In the early 1990s, I wrote the 2G SS7 TCAP (Transaction Capabilities Application Part) protocol for a European telephone exchange maker and enjoyed a protracted struggle with C language, UNIX, a few types of Assembler languages, and a range of European and worldwide standards and recommendations. I also experienced the expansion of digital mobile telephony into Russia and China in the early to mid-1990s and developed software for SMS (Short Message Service) at a time when nobody used the term 'texting' and when adding text messaging to mobile communications was considered by many (including myself) an unpromising idea.[2]

However, over time I started questioning my own engagement with technology. Perhaps a mix of my background in the humanities, my general wariness of the corporate environment, and the political commitment to 'thinking differently' that came from my involvement with the Italian Left throughout the 1990s made me quite conscious of the limits of the merely technical understanding of technology. Thus, this book also stems from my desire to ask different questions about technology from those posed by the technical environment in which I had an opportunity to work. In 2004, when I first began investigating the nature of software from a non-technical point of view, the context of media and cultural studies presented the most interesting academic framework for such an enquiry—although I have also ended up questioning media and cultural studies' approach to technology in the process.

The principal aim of *Software Theory* is to propose a theoretically informed analysis of software which in turn can shed light on the group of media and technologies commonly referred to as 'new' or 'digital'. Taking into account the complexity and shifting meanings of the term 'new technologies', I understand these technologies as sharing one common characteristic: they are all based on software. *Software Theory* takes both 'software' and 'theory' seriously by engaging in a rigorous examination and in a deep questioning of the received concepts of 'technology', 'society', 'culture', 'politics', and even the 'human' that media and cultural studies rely upon. Although in this book 'theory' functions as a synonym for a deconstructive tradition of thought that draws mainly (but not exclusively) on Jacques Derrida's work, it also indicates a broader engagement with Continental philosophy on the question of technology. In fact, I want to argue that media and cultural studies can highly benefit from a productive dialogue with philosophy on the subject of technology. In a way, *Software Theory* uses 'theory' to shed light on software, while at the same time using software to shed light on 'theory', and ultimately on the relationship between technology and knowledge (including academic knowledge). I engage in a close and sustained examination of both technical literature and computer code in order to show how the conceptual system underlying software works and how software has become what it is through a process of constant and unstable self-redefinition. Through such deconstructive reading, I destabilize some of the oppositions currently deployed in the cultural study of software (such as those between writing and code, textuality and materiality, the transcendental and the empirical, the human and the technical, technology and society), and I show how software cannot be fully understood merely as an instrument. Instead, I propose an understanding of software as constitutive of the very concept of the human, which is always already machinic and hence also *in*human. I argue for the importance of thinking software philosophically if we are to think it politically, and ultimately—as ambitious and radical as this conclusion might sound—I suggest that we can only imagine a political future today by thinking it *with* and *through* technology, and therefore also with and through software.

When I started my investigation of what were then referred to as new technologies, the currency of the term 'software' in media and cultural studies did not yet match that of 'new technologies', although its appearance was becoming more and more frequent. A decade later, software has become a fashionable object of study. New academic fields have emerged that concentrate on the cultural and philosophical analysis of software, such as Software Studies, Critical Code Studies, and, at least to a certain extent, Digital Humanities. As Wendy Chun argues in her 2011 book entitled *Pro-*

grammed Visions, software is increasingly considered the focal point of digital media studies because of the perceived need to address the invisibility, ubiquity, and power of digital media.[3] This has led many scholars in the field to wonder whether a global picture of 'digital media' as a concept and a discipline is possible at all. In this light, software has been positioned as a 'minimal characteristic' of digital media.

This growing interest in software can be observed both in the United States and in Europe. It mobilizes scholars from very different disciplines, including social sciences, computer science, and the humanities. Since in *The Language of New Media* Lev Manovich called for 'software studies' as the direction to be taken by new media studies, a number of books have been published that tackle the topic of software—one can think here of volumes such as Alexander Galloway's *Protocol*, Katherine Hayles's *My Mother Was a Computer*, Adrian Mackenzie's *Cutting Code*, Matthew Fuller's *Software Studies*, Wendy Chun's *Programmed Visions*, David Berry's *The Philosophy of Software*, Florian Cramer's *Anti-Media: Ephemera on Speculative Arts*, Nick Montfort and others' *10 PRINT CHR$(205.5+RND(1)); : GOTO 10*, and Lev Manovich's own recent book entitled *Software Takes Command*.[4] Conferences have been held—for instance, the Working Groups on Critical Code Studies organized by the Humanities and Critical Code Studies Laboratory (HaCCS Lab) at the University of Southern California, which began in 2010 and involved more than a hundred academics and professionals from a variety of fields. These initiatives, in turn, have also given rise to electronic publications, blogs, and other forms of cultural debate around software.

Overlapping with software and code studies, the digital humanities embrace 'all those scholarly activities in the humanities that involve writing about digital media and technology as well as being engaged in processes of digital media production, practice and analysis'.[5] In Gary Hall's broad definition, the digital humanities encompass not just the creation of interactive electronic archives and literature, or the production of online databases, wikis, and virtual galleries and museums, but also the development of new media theory and the investigation of how digital technologies reshape teaching and research. In what some have described as a 'computational turn', the humanities increasingly see techniques and methodologies drawn from computer science—image processing, data visualization, network analysis—being used to produce new ways of understanding and approaching humanities texts.

And yet, Hall points out that, however significant, the computational turn does not exhaust the theoretical and methodological approaches encompassed by the digital humanities. Recent debates have highlighted the importance of a critical study of software for this field. One can think here of the 2011 issue of the academic journal *Culture Machine* devoted to 'The Digital Humanities Beyond Computing', as well as of Katherine Hayles's book of 2013 entitled *How We Think*, in which she positions software and code studies as subfields of the digital humanities.[6]

In sum, the cultural study of software appears crucial today inside and outside the university, at an international level and across disciplines. Yet it is not only humanities scholars or software professionals who are taking an interest in the cultural debate on software. Artists have become interested in software as well. A number of code art pieces have been produced since the digital art festival Ars Electronica chose 'Code' as its topic—for instance, Zach Blas's 'Disingenuous Bar', or Lorenz Tunny's 'TransCoder', presented at the Documents Event #1–UCLA in June 2007. Exhibitions have been organized that interrogate the nature of software, such as the 'Open Source Embroidery: Craft + Code' exhibition held in 2008 at the HTTP Gallery in London. While the university and the arts are witnessing the emergence of such a vibrant and diversified new field of investigation, the wider public is also becoming increasingly interested in the cultural discussion of software. This is due, again, to the pervasiveness of software and to the fact that media users are constantly being asked to make decisions about software-based technologies in everyday life (for instance, when having to decide whether to use commercial or free software, or whether to upgrade their computers).

And yet, it seems to me that the meaning of 'software' remains not only shifting but also rather unclear. Therefore, throughout the course of *Software Theory* I keep returning to what I see to be a fundamental question that needs to be posed and re-posed in the present cultural context: 'what is software?'. I argue that this question needs to be dealt with seriously if we want to begin to appreciate the role of technology in contemporary culture and society. In other words, I call for a radical 'demystification' of new technologies through a demystification of software. I also maintain that in order to understand new technologies we need first of all to address the mystery that surrounds their functioning and that affects our comprehension of their relationship to the cultural and social realm.

As anticipated above, this will ultimately involve a radical rethinking of what we mean by 'technology', 'culture', and 'society'.

I am particularly interested in the political significance that our — conscious and unconscious — involvement with technology carries. Therefore, with *Software Theory* I also seek a way to think new technologies politically. More precisely, I argue that the main political problem with new technologies is that they exhibit — in the words of Bernard Stiegler — a 'deep opacity'. As Stiegler maintains, 'we do not immediately understand what is being played out in technics, nor what is being profoundly transformed therein, even though we unceasingly have to make decisions regarding technics, the consequences of which are felt to escape us more and more.'[7] I suggest that, in order to develop political thinking about new technologies, we need to start by tackling their opacity.[8]

However, my aim in *Software Theory* is to propose quite a different approach to the investigation of software from the ones offered by software and code studies so far. Rather than looking at what people *do* with software (that is, at the practices of software producers, users, and consumers), or at what software *does* (that is, at how it works), I want to problematize software *as software*, by 'undoing' the conceptual system on which it relies for its functioning. In fact, I want to show how, to a certain extent, software is always in the process of 'undoing itself'. To be able to expand on this point, let me start from a brief examination of the place of new technologies, and particularly of software, in today's academic debate. We have already seen how new technologies are an important focus of academic reflection, particularly in media and cultural studies. With the formulation 'media and cultural studies' I mean to highlight that the reflection on new technologies is positioned at the intersection of the academic traditions of cultural studies and media studies. Nevertheless, to think that technology has only recently emerged as a significant issue in media and cultural studies would be a mistake. In fact, technology (in its broadest sense) has been present in media and cultural studies from the start, as its constitutive concept. The intertwining between the concepts of 'medium' and 'technology' dates back to what some define as the 'foundational' debate between Raymond Williams and Marshall McLuhan.[9] In his work, McLuhan was predominantly concerned with the technological nature of the media, while Williams emphasized the fact that technology was always socially and culturally shaped. At the risk of a certain oversimplification, we can say that media and cultural studies has to a large extent been informed by Williams's side of the argument—and has thus focused its attention on the cultural and social formations surrounding technology, while rejecting the ghost of 'technological determinism', and frequently dismissing any overt attention paid to technology itself as 'McLuhanite'.[10] Yet technology has entered the field of media and cultural studies precisely thanks to McLuhan's insistence on its role as an agent of change.

One must be reminded at this point that, in the perspective of media and cultural studies, to study technology 'culturally' means to follow the trajectory of a particular 'technological object' (generally understood as a technological product), and to explore 'how it is represented, what social identities are associated with it, how it is produced and consumed, and what mechanisms regulate its distribution and use.'[11] Such an analysis concentrates on 'meaning', and on the way in which a technological object is made meaningful. Meaning is understood as not arising from the technological object 'itself', but from the way it is represented in the discourses surrounding it. By being brought into meaning, the technological object is constituted as a 'cultural artefact'.[12] Thus, meaning emerges as intrinsic to the definition of culture deployed by media and cultural studies. This is the case in Williams's classical definition of culture as a 'description of a particular way of life', and of cultural analysis as 'the clarification of the meanings and values implicit and explicit in particular ways of life', as well as in a more recent understanding of 'culture' as 'circulation of meanings' (a formulation that takes into account that diverse, often contested meanings are produced, shared and communicated within different social groups, and that they generally reflect the play of powers in society).[13]

When approaching new technologies, media and cultural studies has therefore predominantly focused on the intertwined processes of production, reception, and consumption, that is, on the discourses and practices of new technologies' producers and users. Such an approach has been substantially inherited by the emerging fields of software and code studies. From this perspective, even a technological object as 'mysterious' as software is addressed by asking how it has been made into a significant cultural object. For instance, in his book of 2006 entitled *Cutting Code*, Adrian Mackenzie investigates 'software as a social object and process' and demonstrates its relevance as a topic of study essentially by examining the social and cultural formations that surround it.[14] Even though one of the commonly declared aims of software and code studies is to address 'software (or code) itself'—or, in Matthew Fuller's words, 'the stuff of software'[15]—it seems to me that these fields constantly alternate between searching for social and cultural meanings in software on the one hand, and offering technical expositions of 'how software works' on the other.

This, indeed, is the approach adopted by Manovich's aforementioned call for software studies in *The Language of New Media*. Manovich clarifies: 'to understand the logic of new media we need to turn to computer science.'[16] According to Manovich, since contemporary media have become programmable, computer science can provide media scholars with the terminology and the conceptual framework to make sense of them: '[f]rom media studies, we move to something which can be called software studies; from media theory—to software theory.'[17] This 'turn to software', which has been positioned by many as the seminal moment of software and code studies, can hardly be seen as uncontroversial. For instance, again following Gary Hall, one could ask: is computer science really that well equipped to understand itself and its own founding object, let alone help media studies (or, for that matter, the humanities) in understanding their own relation with computing?[18]

In fact, in his recent book, *Software Takes Command*, Manovich acknowledges how this 'turn to software' risks positing computer science as an absolute truth and advances the proposition that 'computer science is itself part of culture' and that 'Software Studies has to investigate the role of software in contemporary culture, and the cultural and social forces that are shaping the development of software itself.'[19] Ultimately, Manovich recommends that software studies focus on software as a cultural and social object—that is, as 'a layer that permeates all areas of contemporary societies', a 'metamedium' that has replaced most other media technologies that emerged in the nineteenth and twentieth centuries.[20] For Manovich, understanding software is a mandatory condition of being able to understand contemporary techniques of communication, interaction, memory, and control. And yet, to understand software we need to understand its history and the social and cultural conditions that have shaped its development.[21]

This ambivalence—one could even say, this circular movement—between the technical and the social seems to me to characterize the whole field of software and code studies. Individual academic contributions can choose to privilege one term over the other, but ultimately they turn to the technical in order to explain the social, and then revert to the social in order to explain the technical. Representative of a stronger emphasis on the technical aspects of new technologies are studies such as Nick Montfort and Ian Bogost's 2009 book on the Atari video computer system, entitled *Racing the Beam*, which has been positioned as the inaugural text of platform studies. Overlapping with software studies, platform studies pays particular attention to the specific hardware and software systems on which 'expressive computing' (another name for Manovich's 'cultural software'—that is, software applications for video games, digital art, and electronic literature) is based. Montfort and Bogost recommend 'delving into the code' and giving 'serious and in-depth consideration [to] circuits, chips, peripherals and how they are integrated and used' in order to make sense of new media.[22] Also intent on the functional unpacking of technology is the classical work of Alexander Galloway, *Protocol* (2004), which offers a welcome counterargument to the widespread discursive use of digital networks as metaphors of horizontal connectivity and intrinsic political empowerment. By unpacking the hierarchical workings of 'real' networks, which are functionally based on layers of software that 'wrap up' the contents of communication, Galloway also unpacks and critiques the rhetoric of freedom that surrounds digital networks. For instance, he demonstrates how protocols incorporate ('embed') censorship and other mechanisms of distributed control, and how these technical characteristics embody a logic of governmentality which he ultimately associates with the military origins of the Internet. 'Distributed networks', Galloway states, 'are native to Deleuze's control society.'[23] In sum, for Galloway, discovering the control mechanisms which found the technical functioning of networks is a way to complicate the political discourse on networks inside and outside the academy. Although Galloway's approach remains very important and politically meaningful for the cultural study of software, it also runs the risk of positioning the technical as the ultimate truth, as showing what 'truly' lies behind any discourse on technology (celebratory or otherwise). In the end, this approach rests on a strategy that draws on technical explanation as a way to reach a 'deeper' understanding of technology—an understanding supposedly located 'behind' or 'beyond' the social.

This 'quest for depth' in the analysis of software, which positions software as the truth of new technologies while also intertwining software with issues of 'power' (or 'biopower'), 'control' and 'governmentality', is also present in more socially orientated studies of software, such as David Berry's 2011 book, *The Philosophy of Software*. For Berry, in order to understand the way in which one experiences software today and to develop a 'phenomenology of software', software studies need to 'unpack' software, to understand what it 'really does', to reach a technical grasp of it.[24] Yet, again, this process of unpacking leads to the analysis of the practices of production, usage, and consumption that shape and are shaped by software. Importantly, according to Berry, such a deeper understanding of software—which, for instance, unmasks the decisions made by big corporations such as Google with regard to their software—is ultimately an emancipatory practice, because it leads individuals to better-informed political decisions. An analogous emancipatory aim characterizes Robert Kitchin and Martin Dodge's book, *Code/Space* (2011), which investigates the way in which software shapes social space by enabling 'forms of automation, the monitoring and controlling of systems from a distance, the reconfiguring and rejuvenation of established industries, the development of new forms of labor practices and paid work, the reorganization and recombination of social and economic formations at different scales' and many other innovations.[25] Kitchin and Dodge develop the idea of 'automated management' in order to explore how software enables a whole new range of movements and transactions across space while also automatically and systematically capturing and storing data on these transactions, thus bringing about new opportunities for personal and collective empowerment which are also new possibilities for regulation and control.

The intertwining between digital technologies and power is also at the core of the articulated analysis of software offered by Wendy Chun in her recent book, *Programmed Visions* (2011). Chun argues that computers, and particularly software, 'embody a certain logic of governing or steering through the increasingly complex world around us', which she calls 'the logic of programmability'.[26] Programmability gives users a sense of empowerment by ultimately incorporating them into the logic of a machine that they do not fully understand but feel somewhat in control of. In other words, we feel empowered by software because we feel in control of something we do not have a grasp of. In fact, Chun argues that our fascination with software resides precisely in the fact that we do not understand it. Instead, we view it as something hidden, obscure, difficult to pin down, and generally invisible, which nevertheless generates visible effects in the world (for instance, lights flickering on a screen or the functioning of telecommunication networks). Rather than aiming at dispelling the mysteriousness of software, she analyzes how software functions as a 'metaphor of metaphors' in a number of discourses and fields of knowledge precisely *because of* its relative obscurity. She writes: '[Software's] combination of what can be seen and not seen, can be known and not known—its separation of inter-face from algorithm; software from hardware—makes it a powerful metaphor for everything we believe is invisible yet generates visible effects, from genetics to the invisible hand of the market; from ideol-ogy to culture.'[27] Although critical of those approaches of software studies that view knowing software as 'a form of enlightenment', in her own analysis Chun still combines historical-cultural narratives (for instance, the history of programming languages) with technical explanations (for instance, the exposition of how digital circuits work) in order to complicate her cultural account of technology and particularly of software.

This political commitment with the analysis of software has been taken further by a number of studies on software which draw on neomaterialism, media archaeology, and object-oriented philoso-phy.[28] In their quest for thinking software as a material entity or process which spreads into economics, politics, and—again—the whole logic of control society as an 'immanent' force understood in a Simondonian and Deleuzian sense, these studies tend to 'ontolo-gize' software as the condition of possibility of contemporary life, and to privilege software studies as a master discourse capable of making visible the foundational but otherwise invisible, hidden, embedded, off-shored, or forgotten nature of software and code. And yet, one could ask: To what extent can software and code be privileged in this respect? On what basis can they be said to consti-tute the conditions for revealing the truth of human life or society?[29]

So, to recap, not only is the number of existing studies of soft-ware still relatively limited, but these studies also give an account of software that is based on the analysis of the processes of software production, reception, and consumption combined with the techni-cal exposition of how software 'really' works (either as a technical artefact or as the all-pervasive logic of control societies). Although I recognize that the above perspective remains absolutely relevant to the political and cultural study of technology, I suggest that this approach should be supplemented by an alternative, or I would even hesitantly say more 'theoretical', and yet more 'direct', investi-gation of software—although I will raise questions for both these notions of 'theory' and 'directness' later on. As I have suggested above, in order to understand the role that new technologies play in our lives and the world as a whole we do need to shift the focus of analysis from the practices and discourses concerning them to a thorough investigation of how new technologies work, and, in par-ticular, of how software works and of what it does. And yet, in

Software Theory I propose an investigation of software that takes historical-cultural and technical narratives as a starting point, rather than as a final one, and I suggest that these narratives should be problematized rather than treated as explanatory. Such an investigation of software will help me to problematize the intertwining of the technical and the social aspects of software which currently preoccupies Software Studies. At the same time, I will refrain from making any sweeping ontological claims about what software *is* and will instead engage in a critical examination of the alleged relations between 'software'/'code' on the one hand and 'ontology' and 'materiality' on the other hand, as conceptualized by cultural theorists of computation such as Katherine Hayles. This is why earlier on I suggested that the title of this book, *Software Theory*, hints at a different engagement with theory from what Lev Manovich had in mind.[30] Let me now explain how such an investigation of software can be undertaken.

By arguing for the importance of such an approach, I do not mean that a 'direct observation' of software is possible. I am well aware that any relationship we can entertain with software is always mediated, and that software might well be 'unobservable'. In fact, I intend to take away all the implications of 'directness' that the concept of 'demystifying' or 'engaging with' software may bring with it. I am particularly aware that software has never been univocally defined by any disciplinary field (including technical ones) and that it takes different forms in different contexts. For instance, a computer program written in a programming language and printed on a piece of paper is software. When such a program is executed by a computer machine, it is no longer visible, although it might remain accessible through changes in the status of the machine (such as the blinking of lights or the flowing of characters on a screen)—and it is still defined as software. Chun has shown how one particular form of software, the source code (that is, computer programmes written in high-level programming languages), became the most representative form of software in the 1970s in order to make software copyrightable, and thus profitable, in the context of the software industry—a process that she names the 'fetishization' of source code, or the emergence of software as a commodity and a 'thing'.[31] Different software studies scholars rely on different definitions of software, from the broadest possible ones to more restricted notions of 'cultural' or 'expressive' software.[32] This obvious difficulty in finding a point of departure when studying software—a difficulty shared by computer science—hints not just at the elusiveness and 'opacity' of software but most importantly at the mediated nature of our access to it.

In *Software Theory* I investigate different forms taken by software, from so-called specifications to source code and run-time code. However, I want to start from a rather widely accepted definition of software as the totality of all computer programs as well as all the written texts related to computer programs. This definition constitutes the conceptual foundation of software engineering, a technical discipline born in the late 1960s to help programmers design software cost-effectively. Software engineering describes software development as an advanced writing technique that translates a text or a group of texts written in natural languages (namely, the requirements specifications of the software 'system') into a binary text or group of texts (the executable computer programs), through a step-by-step process of gradual refinement. As Ian Sommerville explains, 'software engineers model parts of the real world in software. These models are large, abstract and complex so they must be made visible in documents such as system designs, user manuals, and so on. Producing these documents is as much part of the software engineering process as programming.'[33]

This formulation shows that 'software' does not only mean 'computer programs'. A comprehensive definition of software also includes the whole of technical literature related to computer programs, including methodological studies on how to design computer programs—that is, including software engineering literature itself. The essential move that such an inclusive definition allows me to make consists in transforming the problem of engaging with software into the problem of reading it. In *Software Theory* I will therefore ask to what extent and in what way software can be described as legible. Moreover, since software engineering is concerned with the methodologies for writing software, I will also ask to what extent and in what way software can actually be seen as a form of writing. Such a reformulation will enable me to take the textual nature of software seriously. In this context, concepts such as 'reading', 'writing', 'document', and 'text' are no mere metaphors. Rather, they are software engineering's privileged mode of dealing with software as a technical object. It could even be argued—as I shall throughout this book—that in the discipline of software engineering, software's technicity is dealt with as a form of writing.

As a first step, it is important to notice that, in order to investigate software's readability and to attempt to read it, the idea of reading itself needs to be interrogated. In fact, if we accept that software presents itself as a distinctive form of writing, we need to be aware that it consequently invites a distinctive form of reading. But to read software as conforming to the strategies it enforces upon its reader would mean to read it the way a computer professional would, that is, in order to make it function *as software*. I argue that reading software on its own terms is not equal to reading it functionally. For this reason, I develop a strategy for reading software by drawing on Jacques Derrida's concept of 'deconstruction'. However controversial and uncertain a definition of 'deconstruction' might be, I am essentially taking it up here as a way for stepping outside of a conceptual system while simultaneously continuing to use its concepts and demonstrating their limitations.[34] 'Deconstruction' in this sense aims at 'undoing, decomposing, desedimenting' a conceptual system, not in order to destroy it but in order to understand how it has been constituted.[35] According to Derrida, in every conceptual system we can detect a concept that is actually unthinkable within the conceptual structure of the system itself—therefore, it has to be excluded by the system, or, rather, it must remain unthought to allow the system to exist. A deconstructive reading of software therefore asks: what is it that has to remain unthought within the conceptual structure of software?[36] In Derrida's words, such a reading looks for a point of 'opacity', for a concept that escapes the foundations of the system in which it is nevertheless located and for which it remains unthinkable. It looks for a point where the conceptual system that constitutes software 'undoes itself'.[37] For this reason, a deconstructive reading of software is the opposite of a functional reading. For a computer professional, the point where the system 'undoes itself' is a malfunction, something that needs to be fixed. From the perspective of deconstruction, in turn, it is a point of revelation, one in which the conceptual system underlying the software is clarified. Actually, I want to suggest that Derrida's point of 'opacity' is also simultaneously the locus where Stiegler's 'opacity' disappears, that is where technology allows us to see how it has been constituted. Being able to put into question at a fundamental level the premises on which a given conception of technology rests would prove particularly important when making decisions about it, and would expand our capacity for thinking and using technology politically, not just instrumentally.[38]

Let me consider briefly some of the consequences that this examination of software might have for the way in which media and cultural studies deals with new technologies. We have already seen that the issue of technology has been present in media and cultural

studies from the very beginning, and that the debate around technology has contributed to defining the methodological orientation
of the field. For this reason, it is quite understandable that rethinking technology would entail a rethinking of media and cultural
studies' distinctive features and boundaries. A deconstructive reading of software will enable us to do more than just uncover the
conceptual presuppositions that preside over the constitution of
software itself. In fact, such an investigation will have a much larger
influence on our way of conceptualizing what counts as 'academic
knowledge'. To understand this point better, not only must one be
reminded that new technologies change the form of academic
knowledge through new practices of scholarly communication and
publication as well as shifting its focus, so that the study of new
technologies has eventually become a 'legitimate' area of academic
research. Furthermore, as Gary Hall points out, new technologies
change *the very nature and content* of academic knowledge.[39] In a
famous passage, Derrida wondered about the influence of specific
technologies of communication (such as print media and postal services) on the field of psychoanalysis by asking 'what if Freud had
had e-mail?'[40] If we acknowledge that available technology has a
formative influence on the construction of knowledge, then a reflection on new technologies implies a reflection on the nature of academic knowledge itself. But, as Hall maintains, paradoxically 'we
cannot rely merely on the modern "disciplinary" methods and
frameworks of knowledge in order to think and interpret the trans-
formative effect new technology is having on our culture, since it is
precisely these methods and frameworks that new technology requires us to rethink.'[41] According to Hall, cultural studies is the
ideal starting point for a study of new technologies, precisely because of its open and unfixed identity as a field. A critical attitude
towards the concept of disciplinarity has characterized cultural
studies from the start. Such a critical attitude informs cultural studies' own disciplinarity, its own academic institutionalization.[42] Yet
Hall argues that cultural studies has not always been up to such
self-critique, since very often it has limited itself to an 'interdiscipli-
narity' attitude understood only as an incorporation of heterogeneous elements from various disciplines—what has been called the
'pick'n'mix' approach of cultural studies—but not as a thorough
questioning of the structure of disciplinarity itself. He therefore suggests that cultural studies should pursue a deeper self-reflexivity, in
order to keep its own disciplinarity and commitment open. This
self-reflexivity would be enabled by the establishment of a produc-

tive relationship between cultural studies and deconstruction.[43] The latter is understood here, first of all, as a problematizing reading that would permanently question some of the fundamental premises of cultural studies itself. Thus, cultural studies would remain acutely aware of the influence that the university, as a political and institutional structure, exercises on the production of knowledge (namely, by constituting and regulating the competences and practices of cultural studies practitioners). It is precisely in this awareness, according to Hall, that the political significance of cultural studies resides. Given that media and cultural studies is a field which is particularly attentive to the influences of the academic institution on knowledge production, and considering the central role played by technology in the constitution of media and cultural studies, as well as its potential to change the whole framework of this (already self-reflexive) disciplinary field, I want to argue here that a rethinking of technology based upon a deconstructive reading of software needs to entail a reflection on the theoretical premises of the methods and frameworks of academic knowledge.

To conclude, in *Software Theory* I propose a reconceptualization of new technologies, that is of technologies based on software, through a close, even intimate, engagement with software itself, rather than through an analysis of how new technologies are produced, consumed, represented, and talked about. To what extent and in what way such intimacy can be achieved and how software can be made available for examination are the main questions addressed in this book. Taking into account possible difficulties resulting from the working of mediation in our engagement with technology as well as technology's opacity and its constitutive, if unacknowledged, role in the formation of both the human and academic knowledge, I want to argue via close readings of selected software practices, inscriptions, and events that such an engagement is essential if we are to take new technologies seriously, and to think them in a way that affects—and that does not separate—cultural understanding and political practice.

In Chapter 1, I suggest that the problem of digital technologies, and of the kind of knowledge that can be produced about them, cannot be addressed without radically reconsidering what we mean by 'knowledge' in relation to 'technology' in a broader sense. I argue that, as a preliminary step, the received concepts of technology—that is, the ways in which technology has been understood primarily by the Western philosophical tradition—need to be put into question. I outline two traditions of philosophical thought about technology: the dominant Aristotelian conception of technology as an instrument, and an alternative line of thought based on the idea of the 'originary technicity' of the human being. Subsequently, I draw on the work of thinkers belonging to the latter tradition (Martin Heidegger, Bernard Stiegler, André Leroi-Gourhan, Jacques Derrida) to develop a different way of thinking about technology, one that will ultimately prove more productive for my investigation of software.

I argue for a radical rethinking of the conceptual framework of instrumentality if an understanding of technology, and specifically of software, is to be made possible. A pivotal role is played here by the idea of linearization, which also belongs to the tradition of originary technicity: it was developed by Leroi-Gourhan and subsequently reread by Derrida in *Of Grammatology* in the context of his own reflections on writing.[44] By becoming a means for the phonetic recording of speech, writing became a technology—it was placed at the level of a tool, or of 'technology' in its instrumental sense. Derrida relates the emergence of phonetic writing to a linear understanding of time and history, thus emphasizing the relationship between technology, language, and time. Ultimately, I draw on Stiegler's thought on technology and on his own rereading of Derrida's work in order to call for a concrete analysis of historically specific technologies—including software—while keeping open the significance of such an analysis for a radical rethinking of the relationship between technology and the human.[45]

To what extent and in what way such an investigation of software can be pursued is the focus of Chapter 2, which deals with the concept of 'writing' in relation to 'language' and 'code' more closely and examines the usefulness of these concepts for the understanding of software. I draw mainly on Hayles's understanding of the 'regime of code' as opposed to the regimes 'of speech' and 'of writing', and on her suggestion that writing and code are intertwined in software. Nevertheless I question her assumption that software, as a technical object, is 'beyond metaphysics', and I propose a different reading of Derrida that views his notion of 'writing', as well as of deconstruction, as a promising theory of technology capable of inspiring innovative strategies for the analysis of software. I argue that, since his earliest works (in particular, *Of Grammatology*), Derrida has defined writing as a material practice. For Derrida, we need to have an understanding of writing in order to grasp the meaning of orality—not because writing historically existed before language, but because we must have a sense of the permanence of a linguistic mark in order to recognize it. Ultimately, a sense of writing is necessary for signification to take place. In other words, language itself is material; it needs materiality (or rather, it needs the possibility of an 'inscription') to function as language. Software itself can function only through materiality, because software is (also) code, and materiality is what constitutes signs (and therefore codes). But if every bit of code is material, and if the material structure of the mark is at work everywhere, how are we supposed to study software as a historically specific technology? I argue that the historical specificity of software resides in its constitution through the continuous undoing and redoing of the boundaries between 'software' itself, 'writing', and 'code'. In order to do so I take a critical approach to Stiegler's distinction between 'technics' and 'mnemotechnics', which becomes untenable when applied to software. Ultimately, I argue that software remains to some extent implicated with the instrumental understanding of technology while at the same time being able to escape it by bringing forth unexpected consequences

In Chapter 3, I show how, in the late 1960s, 'software' emerged precisely as a specific construct in relation to 'writing' and 'code' in the discourses and practices of software engineering. I explain how the discipline of software engineering emerges as a strategy for the industrialization of the production of software at the end of the 1960s through an attempt to control the constitutive fallibility of software-based technology. Such fallibility—that is, the unexpected consequences inherent in software—is dealt with through the organization and linearization of the time of software development. Software engineering also understands software as the 'solution' to preexistent 'problems' or 'needs' present in society, therefore advancing an instrumental understanding of software. However, both the linearization of time and the understanding of software as a tool are continuously undone by the unexpected consequences brought about by software—which must consequently be excluded and controlled in order for software to reach a point of stability. At the same time, such unexpected consequences remain necessary to software's development.

In Chapter 4, I turn to the evolution of software engineering in the 1970s and 1980s into a discipline for the management of time in software development. I start by examining two of the fundamental texts on the management of time in software development, both written by Frederick Brooks in the mid-1970s and mid-1980s, respectively, in order to show how the mutual co-constitution of 'software', 'writing', and 'code' develops in these texts.[46] The sequencing of time that software engineering proposed in the 1970s and 1980s, which was justified through an Aristotelian distinction between the essential and the accidental (the ideal and the material), ultimately gave rise to an unexpected reorganization of technology according to the open source style of programming in the 1990s. The framework of instrumentality that, as I showed in Chapter 3, emerged from the 'software crisis' of the late 1960s—that is, from the need to control the excessive speed of software growth—was enforced on software production through the 1970s and 1980s, and it enabled the development of software during these decades as well as the unexpected emergence of open source. I investigate a classic of software engineering in the 'open source era', Eric Steven Raymond's 'The Cathedral and the Bazaar', which responds to Brooks's theories from the point of view of the mid-1990s. I show how open source still takes software engineering as its model—a model in which the framework of instrumentality is once again re-enacted.[47] The aim of open source is still to obtain 'usable' software, but 'usability'—that is, a stable version of a software system—is not something that can be scheduled in time. Rather, it is something that happens to the system while it is constantly reinscribed. In sum, in Chapters 3 and 4, I argue that software engineering is characterized by the continuous opening up and reaffirmation of the instrumentality of software and of technology. Furthermore, in Chapter 4, my examination of the political consequences of Linus Torvald's technical decisions when developing Linux allow me to show in what sense politics and technology could and should be thought together.

Finally, in Chapter 5, I leave software engineering and turn to the examination of a number of classical works on programming languages and compilers in order to show how software is constituted as a process of unstable linearization.[48] I explain how software is understood by using concepts derived from structural linguistics—namely language, grammars, and the alphabet—and I discuss the concepts of formal notation, formal grammar, programming language, and compiler. I specifically concentrate on the process of 'compiling'—that is, the process through which a program is made executable through its 'translation' from a high-level programming language into machine code. I show how, in the theory of programming languages, the instrumentalization of technology and the linearization of programming languages go hand in hand, and how software exceeds both. In fact, software is defined as a process of substitution—or reinscription—of symbols, which always involves iteration. The theory of programming languages is an attempt to manage iteration through linearization—an attempt ultimately destined to fail, since iteration is always open to variation. Actually, iteration is a constitutive characteristic of software, just as fallibility, and the capacity for generating unforeseen consequences, is constitutive of technology. Every time technology brings about unexpected consequences, it is fundamental to decide whether this is a malfunction that needs to be fixed or an acceptable variation that can be integrated into the technological system, or even an unforeseen anomaly that will radically change the technological system for ever. Ultimately such decisions cannot be made purely on technical grounds—they are, instead, of a political nature.

Taking my clue from the analysis of a number of media examples, in the Conclusions of *Software Theory* I explain how obscuring the incalculability of technology leads to setting up an opposition between risk and control and ultimately to the authoritarian resolution of every dilemma regarding technology. Once again I emphasize that technology (as well as the conceptual system on which it is based) can only be problematized from within, and how such a problematization needs to be a creative and politically meaningful process. Even more importantly, technology—as well as software—needs to be studied in its singularity. In particular, the opacity of software (which is the problem we started fom) cannot be dispelled merely through an analysis of what software 'really is'—for instance, by saying that software is 'really' just hardware. Rather, one must acknowledge that software is always both conceptualized according to a metaphysical framework and capable of escaping it, that it is instrumental and generating unforeseen consequences, that it is both a risk and an opportunity (or, in Derridean terms, a *pharmakon*). If the unexpected is always implicit in technology, and the potential of technology for generating the unexpected needs to be unleashed in order for technology to function, every choice we make with regard to technology always implies an assumption of responsibility for the unforeseeable. Technology will never be calculable—and yet decisions must be made. The only way to make politically informed decisions about technology is not to obscure such uncalculability. By opening up new possibilities and foreclosing others, our decisions about technology also affect our future. Thus, thinking politics with technology becomes part of the process of the reinvention of the political in our technicized and globalized world. Rethinking technology *is* a form of imagining our political future.

NOTES

1. GSM was originally a European project. In 1982, the European Conference of Postal and Telecommunication Administration (CEPT) instituted the Groupe Spécial Mobile (GSM) to develop a standard for a mobile telephone system that could be used across Europe. In 1989, the responsibility for GSM was transferred to the European Telecommunications Standards Institute (ETSI). GSM was resignified as the English acronym for Global System for Mobile Communications, and Phase One of the GSM specifications were published in 1990. The world first public GSM call was made on 1 July 1991 in a city park in Helsinki, Finland, an event which is now considered the birthday of second-generation mobile telephony—the first generation of mobile telephony to be completely digital. In the early 1990s, Phase Two of GSM was designed and launched, and GSM rapidly became the worldwide standard for digital mobile telephony. Two more generations of mobile telecommunications followed: 3G (which was standardized by ITU and reached widespread coverage in the mid-2000s) and the current 4G, which provides enhanced mobile data services, including ultra-broadband Internet access, to a number of devices such as smartphones, laptops, and tablets. See, for instance, Juha Korhonen, *Introduction to 4G Mobile Communications* (Boston: Artech House, 2014).

2. TCAP is a 2G digital signaling system that enables communication between different parts of digital networks, such as telephone switching centres and databases. SMS is a communication service standardized as part of the GSM network (the first definition of SMS is to be found in the GSM standards as early as 1985), which allows for the exchange of short text messages between mobile telephones. The SMS service was developed and commercialized in the early 1990s. Today SMS text messaging is one of the most widely used data application in the world.

3. Wendy Hui Kyong Chun, *Programmed Visions: Software and Memory* (Cambridge, MA and London: MIT Press, 2011), 1.

4. Lev Manovich, *The Language of New Media* (Cambridge, MA and London: MIT Press, 2001); Alexander Galloway, *Protocol: How Control Exists after Decentralization* (Cambridge, MA: MIT Press, 2004); Katherine N. Hayles, *My Mother Was a Computer: Digital Subjects and Literary Texts* (Chicago: University of Chicago Press, 2005); Adrian Mackenzie, *Cutting Code: Software and Sociality* (New York and Oxford: Peter Lang, 2006); Matthew Fuller, ed., *Software Studies. A Lexicon* (Cambridge, MA and London: MIT Press, 2008); Chun, *Programmed Visions*; David M. Berry, *The Philosophy of Software: Code and Mediation in the Digital Age* (Basingstoke: Palgrave Macmillan, 2011); Florian Cramer, *Anti-Media: Ephemera on Speculative Arts* (Rotterdam: NAi Publishers and Institute of Network Cultures, 2013); Nick Montfort, Patsy Baudoin, John Bell, Ian Bogost, Jeremy Douglass, Mark C. Marino, Michael Mateas, Casey Reas, Mark Sample, and Noah Vawter, *10 PRINT CHR$(205.5+RND(1)); : GOTO 10* (Berkeley: MIT Press, 2013); Lev Manovich, *Software Takes Command: Extending the Language of New Media* (New York and London: Bloomsbury, 2013).

5. Gary Hall, "The Digital Humanities beyond Computing: A Postscript," *Culture Machine* 12 (2011), www.culturemachine.net/index.php/cm/article/download/441/471.

6. "The Digital Humanities beyond Computing," [Special Issue], ed. Federica Frabetti, *Culture Machine* 12 (2011); Katherine N. Hayles, *How We Think: Digital Media and Contemporary Technogenesis* (Chicago: University of Chicago Press, 2012).

7. Bernard Stiegler, *Technics and Time, 1: The Fault of Epimetheus* (Stanford, CA: Stanford University Press, 1998), 21.

8. Alfred Gell develops an interesting reflection on the relations between art, technology, and magic (Alfred Gell, "The Technology of Enchantment and the Enchantment of Technology," in *Anthropology, Art, and Aesthetics*, ed. Jeremy Coote and Antony Shelton [Oxford: Clarendon Press, 1992], 40–63). Drawing on the example of Trobriand canoe-boards, Gell argues that the foundation of art is a technically achieved level of excellence that a society misrepresents to itself as a product of magic. Gell views art as a special form of technology and art objects as devices to obtain social consensus. For Gell 'the power of art objects stems from the technical processes they objectively embody: the *technology of enchantment* is founded on the *enchantment of technology*' (Gell, "The Technology of Enchantment," 44). The magical prowess, which is supposed to have entered the making of the art object, depends on the level of cultural understanding that surrounds it. The same can be said of technology: 'the enchantment of technology is the power that technical processes have of casting a spell over us so that we see the real world in an enchanted form' (Gell, "The Technology of Enchantment," 44). This is what Gell names the 'halo effect of technical difficulty' (Gell, "The Technology of Enchantment," 48; Christopher Pinney and Nicholas Thomas, eds., *Beyond Aesthetics: Art and the Technologies of Enchantment*, [Oxford and New York: Berg, 2001]). He argues that art objects are made valuable precisely by virtue of the 'intellectual resistance' they offer to the viewer (Gell, "The Technology of Enchantment," 49) and that 'technical virtuosity is intrinsic to the efficacy of works of art in their social context' because it creates an asymmetry in the relationship between the artist and the spectator (48–52). Gell's reflection contributes to show the cultural character of the sense of enchantment that surrounds technology. Furthermore, by suggesting that such sense of enchantment has a role in the creation of social consensus, it provides evidence to the fact that any attempt to change the processes through which we understand technology has significant political consequences.

9. Martin Lister, Jon Dovey, Seth Giddings, Iain Grant, and Kelly Kieran, *New Media: A Critical Introduction* (London and New York: Routledge, 2003).

10. Lister et al., *New Media*, 73.

11. Paul DuGay, Stuart Hall, Linda Janes, Hugh Mackay, and Keith Negus, *Doing Cultural Studies: The Story of the Sony Walkman* (London: Sage/The Open University, 1997), 3.

12. DuGay et al., *Doing Cultural Studies*, 10.

13. Raymond Williams, *The Long Revolution* (Harmondsworth: Penguin, 1961), 57. See also Stuart Hall, ed., *Representation: Cultural Representations and Signifying Practices* (London: Sage/The Open University, 1997); DuGay et al., *Doing Cultural Studies*.

14. Mackenzie, *Cutting Code*, 1.

15. Fuller, ed., *Software Studies*, 1.

16. Manovich, *The Language of New Media*, 48.

17. Manovich, *The Language of New Media*, 48.

18. Hall, "The Digital Humanities beyond Computing," 3.

19. Manovich, *Software Takes Command*, 10.

20. Manovich, *Software Takes Command*, 16.

21. Manovich, *Software Takes Command*, 4.

22. Nick Montfort and Ian Bogost, *Racing the Beam: The Atari Video Computer System* (Cambridge, MA and London: MIT Press, 2009), 2.

23. Galloway, *Protocol*, 11.

24. Berry, *The Philosophy of Software*, 9.

25. Rob Kitchin and Martin Dodge, *Code/Space: Software and Everyday Life*, Cambridge (MA and London: MIT Press, 2011), 9.

26. Chun, *Programmed Visions*, 9.

27. Chun, *Programmed Visions*, 17.

28. Matthew Fuller and Tony D. Sampson, eds., *The Spam Book: On Viruses, Porn, and Other Anomalies from the Dark Side of Digital Culture* (Cresskill: Hampton Press, 2009); Matthew Fuller and Andrew Goffey, eds., *Evil Media* (Cambridge, MA and London: MIT Press, 2012); Jussi Parikka, "New Materialism as Media Theory: Medianatures and Dirty Matter," *Communication and Critical/ Cultural Studies* 9, no. 1 (2012): 95–100.

29. I would like to thank Gary Hall for pointing this out to me.

30. Manovich, *The Language of New Media*, 48.

31. Chun, *Programmed Visions*, 20.

32. Cf. Berry, *The Philosophy of Software*; Montfort and Bogost, *Racing the Beam*; Manovich, *Software Takes Command*.

33. Ian Sommerville, *Software Engineering* (Boston: Addison-Wesley, 2011), 4.

34. Cf. Jacques Derrida, *Writing and Difference* (London: Routledge, 1980).

35. In "Structure, Sign, and Play in the Discourse of the Human Sciences", while reminding us that his concept of 'deconstruction' was developed in dialogue with structuralist thought, Derrida speaks of 'structure' rather than of conceptual systems, or of systems of thought (Derrida, *Writing and Difference*). However, 'structure' hints at as complex a formation as, for instance, the ensemble of concepts underlying social sciences, or even the whole of Western philosophy. See also Jacques Derrida, "Letter to a Japanese Friend," in *Derrida and Différance*, ed. Robert Bernasconi and David Wood (Warwick: Parousia Press, 1985), 1–5.

36. I am making an assumption here—namely that software is a conceptual system as much as it is a form of writing and a material object. In fact, the investigation of these multiple modes of existence of software is precisely what is at stake in my book. In the context of the present introduction, and for the sake of clarity, I am concentrating on the effects of a deconstructive reading of a 'structure' understood in quite an abstract sense.

37. According to Derrida, deconstruction is not a methodology, in the sense that it is not a set of immutable rules that can be applied to any object of analysis—because the very concepts of 'rule', of 'object', and of 'subject' of analysis, themselves belong to a conceptual system (broadly speaking, they belong to the Western tradition of thought), and therefore are subject to deconstruction too. As a result, 'deconstruction' is something that 'happens' within a conceptual system, rather than a methodology. It can be said that any conceptual system is always in deconstruction, because it unavoidably reaches a point where it unties or disassembles its own presuppositions. On the other hand, since it is perfectly possible to remain oblivious to the permanent occurrence of deconstruction, there is a need for us to actively 'perform' it, that is, to make its permanent occurrence visible. In this sense deconstruction is also a productive, creative process.

38. This methodology for reading software sets itself apart from the general call to 'reading code' advanced by scholars in the field of critical code studies, such as the analyses of source code proposed by Mark Marino. Critical code studies (CCS) 'emerged in 2006 as a set of methodologies that sought to apply humanities-style hermeneutics to the interpretation of the extrafunctional significance of computer source code. . . . The goal of the study is to examine the digitally born artifact through the entry point of the code and to engage the code in an intensive close reading following the models of media archaeology,

semiotic analysis, and cultural studies' (Mark Marino, "Reading Exquisite Code: Critical Code Studies of Literature," in *Comparative Textual Media: Transforming the Humanities in the Postprint Era*, ed. Katherine N. Hayles and Jessica Pressman [Minneapolis and London: University of Minnesota Press, 2013], 283). For instance, in his article 'Disrupting Heteronormative Codes: When Cylons in Slash Goggles Ogle AnnaKournikova', Marino interprets a mouse click as the inscription of normative heterosexuality in the source code of a computer virus (Mark Marino, "Disrupting Heteronormative Codes: When Cylons in Slash Goggles Ogle AnnaKournikova," in *Digital Arts and Culture Proceedings* [University of California Irvine, 2009], http://escholarship.org/uc/item/09q9m0kn). In *Software Theory*, I propose a more radical problematization of code's own presuppositions—that is, those presuppositions that make code work *as code*.

39. Gary Hall, *Culture in Bits: The Monstrous Future of Theory* (London and New York: Continuum, 2002); Gary Hall, *Digitize This Book! The Politics of New Media, or Why We Need Open Access Now* (Minneapolis: University of Minnesota Press, 2008).

40. Jacques Derrida, *Archive Fever: A Freudian Impression* (Chicago: University of Chicago Press, 1996).

41. G. Hall, *Culture in Bits*, 128.

42. G. Hall, *Culture in Bits*, 115. For the scope of the present Introduction, I assume Hall's term 'cultural studies' as roughly equivalent to what I have previously named 'media and cultural studies', since this passage refers to a constitutive debate around the field's conceptual framework.

43. Cf. Gary Hall and Clare Birchall, eds., *New Cultural Studies: Adventures in Theory* (Edinburgh: Edinburgh University Press, 2006).

44. Jacques Derrida, *Of Grammatology* (Baltimore: The Johns Hopkins University Press, 1976); André Leroi-Gourhan, *Gesture and Speech* (Cambridge, MA: MIT Press, 1993).

45. Cf. Bernard Stiegler, *Technics and Time 1–3* (Stanford, CA: Stanford University Press, 1998–2011).

46. Frederick. P. Brooks, *The Mythical Man-Month: Essays on Software Engineering* (Reading, MA: Addison-Wesley, 1995); Frederick. P. Brooks, "No Silver Bullet: Essence and Accidents of Software Engineering," *IEEE Computer* 20, no. 4 (1987): 10–19.

47. Eric S. Raymond, *The Cathedral and the Bazaar: Musings on Linux and Open Source by an Accidental Revolutionary* (Cambridge, MA: O'Reilly, 2001b).

48. John E. Hopcroft and Jeffrey D. Ullman, *Formal Languages and Their Relation to Automata* (Reading, MA: Addison-Wesley, 1969); Alfred V. Aho and Jeffrey D. Ullman, *Principles of Compiler Design* (Reading, MA: Addison-Wesley, 1979).

ONE

From Technical Tools to Originary Technicity

The Concept of Technology in Western Philosophy

In this chapter, I suggest that the problem of 'new technologies', and of the kind of knowledge that can be produced about them, cannot be addressed without radically reconsidering what we mean by 'knowledge' in relation to 'technology' in a broader sense. I also argue that, as a preliminary step, the received concepts of technology need to be put into question. By received concepts of technology I mean the ways in which technology has been understood primarily by the Western philosophical tradition. This turn to the philosophical conceptions of technology in the context of media and cultural studies might seem somewhat daring or even misjudged. However, I argue that media and cultural studies can highly benefit from a productive dialogue with philosophy on the subject of technology. In fact, a debate on the relevance of philosophical thought has taken place, from time to time, throughout the history of media and cultural studies. This debate has mainly focused on 'theory', that is, on specific developments in French structuralist and post-structuralist thought, such as semiotics and deconstruction. Nevertheless, I want to point out here that this capacity for questioning its own conceptual framework is precisely what enables media and cultural studies to *think technology* originally and innovatively, and therefore to interrogate what we mean by technology in the first place.

To give but one example, in his famous essay 'Cultural Studies and Its Theoretical Legacies', Stuart Hall takes into consideration the tension between theoretical and political dimensions that for him determines the specificity of cultural studies. He writes:

> Both in the British and the American context, cultural studies has drawn the attention itself, not just because of sometimes dazzling internal theoretical development, but because it holds theoretical and political questions in an ever irresolvable but permanent tension. It constantly allows one to irritate, bother, and disturb the other, without insisting on some final theoretical closure.[1]

According to Hall, the theoretical encounters with structuralism and poststructuralism have forced cultural studies to constantly question itself and to keep its identity open and heterogeneous. And yet what holds the field together is its politically committed nature. Hall comments: '[n]ot that there is one politics already inscribed within it. But there is something *at stake* in cultural studies in a way that I think, and hope, is not exactly true of many other very important intellectual and critical practices.'[2]

In his book of 2002 entitled *Culture in Bits*, Gary Hall observes that, while acknowledging the tension between theory and politics, Stuart Hall is actually inclined to give priority to the latter: therefore politics remains that which limits the destabilizing and decentring effects of theory.[3] Gary Hall takes a much more far-reaching stance by suggesting that theory itself has political relevance in cultural studies. For him, by enabling reflexivity within cultural studies, theory also enables cultural studies to become particularly aware of the influences that the university as a political and institutional structure exerts on the production of knowledge, including knowledge produced within cultural studies. Thinking politically means first of all being attentive to the institutional forces that shape thought itself, such as the constitution and regulation of cultural studies practitioners' competences by the university. Therefore, since the ability to question inherited conceptual frameworks appears to be one of cultural studies' points of strength, and since *Software Theory* aims at 'demystifying' new technologies and developing new forms of knowledge about technology and software within media and cultural studies, I want to argue here that a reexamination of the philosophical conceptions of technology is a convenient starting point for my argument.

Indeed, as Bernard Stiegler remarks, "Western philosophy has always found it rather difficult to think about technology." In the first volume of *Technics and Time*, Stiegler points out how, while the extraordinary technological changes of our age need to be conceptualized and made intelligible as soon as possible, in attempting to achieve this intelligibility one cannot rely on any available account of technology in the Western philosophical tradition: 'At its very origin and up until now, philosophy has repressed technics as an object of thought. Technics is the unthought.'[4] Although later on in his work Stiegler identifies a few exceptions to this philosophical refusal to openly approach technology—namely, the thought of several French philosophers, including Jacques Derrida, and that of Martin Heidegger—he nevertheless points out that philosophical reflection has traditionally pushed technology to its own margins. And yet, a critical evaluation of such reflection shows how the concept of technology has always been tightly connected to the concepts of 'knowledge', 'language', and 'humanity'.

For this reason, in the present chapter, I take into consideration a number of philosophers who have attempted a conceptualization of technology and dealt with the difficulty of producing knowledge about it. I examine both the dominant philosophical conception of technology based on the Aristotelian thought, which substantially reduces technology to a mere instrument, and the work of those thinkers who have distanced themselves from such an instrumental understanding and have instead proposed a view of technology as a fundamental characteristic of human beings. I refer mainly, but not exclusively, to the work of Heidegger, Stiegler, Derrida, and the French palaeontologist André Leroi-Gourhan. The work of all these thinkers shows that philosophy has constituted itself precisely in relation (and in opposition) to technological knowledge, and therefore it points to the need for the radical rethinking of philosophy itself if an understanding of technology is to be made possible.

Tracing a map of the philosophical thought on technology is not an easy task. In order to start exploring this problem, let me follow for a moment the innovative genealogy proposed by Stiegler.[5] Stiegler's position on the relationship between philosophy and technology is quite striking. Although, as we have seen above, he argues for the 'urgency and necessity of an encounter between philosophy and technology', he actually views philosophy as traditionally and constitutively incapable of thinking technology.[6] For him, philosophy has always 'repressed' technology as an object of thought. Even more significantly, from the very beginning Western philosophy has distinguished itself from technology, and has in fact identified itself as *not* technology. It has done so by separating *technê* from *epistêmê*. *Epistêmê* is the Greek word most often translated as 'knowledge', while *technê* is translated as either 'craft' or 'art'.[7] The separation between *technê* and *epistêmê* was rooted in the political arena of fifth-century Athens, and it associated *technê* with the rhetorical skills of the Sophists. As professional rhetoricians, the Sophists were skilled in the construction of political arguments. Their skillfulness (*technê*) was perceived as indifference to establishing truth or, worse, as an attempt to make truth instrumental to power. As such, Sophists' *technê* came to be opposed to true knowledge. Therefore, truth remained the only object of *epistêmê*, which in turn was identified with philosophy. This substantially political move deprived technical knowledge of any value.

The subsequent step in the devaluation of technology was made by Aristotle through his definition of a 'technical being' as something that does not have an own end in itself and that is just a tool used by someone else for their ends.[8] In other words, the exclusion of technology from philosophy has been founded on the concept of instrumentality: technical knowledge has been interpreted as instrumental, and therefore as non-philosophy. To quote Timothy Clark, 'the conception of technology that . . . has dominated Western thought for almost three thousand years' can be synthesized as follows: '[t]he traditional, Aristotelian view is that technology is extrinsic to human nature as a tool which is used to bring about certain ends. Technology is applied science, an instrument of knowledge. The inverse of this conception, now commonly heard, is that the instrument has taken control of its maker, the creation control of its creator (Frankenstein's monster).'[9] Moreover, instrumentality has gained a new importance during the process of the industrialization of the Western world. Accordingly, technology has slowly acquired a new place in philosophical thought. Science has in fact become more and more instrumental (to economy, to war) in the course of the last two centuries, therefore gradually renouncing its character of 'pure' knowledge. At the same time, philosophy has become interested in the 'technicization' of science. As an example of this one can think of Edmund Husserl's work on the arithmeticization of geometry.[10] Importantly, as Stiegler also points out, the Platonic conception of technicization as the loss of memory is still at the basis of Husserl's understanding of algebra.[11] I will come back to Plato's understanding of technology in a moment. For now it is worth remembering that in his dialogue *Phaedrus* Plato famously associates writing, understood as a technique to aid memory, with the loss of true memory, which for Plato is *anamnesis*, or recollection of an ideal truth. From this perspective, which again separates knowledge from technology, writing is devalued because of its instrumentality.[12]

To recapitulate the above argument, the devaluation of technology in Western philosophy goes hand in hand with the devaluation of writing. What I want to argue here is that the relationship established by Stiegler between technology and writing as both excluded by knowledge and encompassed by the concept of instrumentality assumes a particular importance in the context of the study of new technologies. The question to be posed at this point is not just whether an instrumental concept of technology is adequate for an understanding of new technologies, but also: if new technologies exceed and destabilize the concept of instrumentality, do they not also destabilize the concept of writing? And what would the consequences of such a destabilization be for the investigation of software?

In order to develop this point, it is important to examine further the alternative tradition of thought on technology that, again according to Stiegler, starts with Heidegger and is not based on the concept of instrumentality. Clark calls this the tradition of 'originary technicity'—a term he borrows from Richard Beardsworth.[13] This term assumes a paradoxical character only if one remains situated within the instrumental conceptualization of technology: if technology were instrumental, it could not be originary—that is, constitutive of the human. Therefore, the concept of 'originary technicity' resists the utilitarian conception of technology. To clarify what he means by 'originary technicity', Clark refers to the 1992 novel *The Turing Option*, coauthored by Marvin Minsky, a leading theorist in the field of Artificial Intelligence.[14] In order to regain his cognitive capacities after a shooting accident has severely damaged his brain, the protagonist of the novel, Brian Delaney, has a small computer implanted into his skull as a prosthesis. After the surgery he starts reconstructing the knowledge he had before the shooting. The novel shows him trying to catch up with himself through his former notes and getting an intense feeling that the self that wrote those notes in the past is lost forever. Clark uses this story as a brilliant figuration of the fact that no self-consciousness can be reached without technology. He comments: 'Delaney's experience in *The Turing Option* is only different in degree from the normal working of the mind from minute to minute. . . . No thinking—no interiority of the psyche—can be conceived apart from technics in the guise of systems of signs which it may seem to employ but which are a condition of its own identity.'[15] Here 'technics' is not understood in terms of massive engineering works but as 'the subtler intimacy of the relation of technology to human thinking', and especially as 'the intimacy between technology and language'.[16] Such an understanding of technology ostensibly draws on Heidegger's thought, as well as on Derrida's and Stiegler's.

As Stiegler points out, it is Heiddeger's understanding of technology that offers the first opportunity to rethink instrumentality and consequently the relationship between technology and knowledge.[17] Famously, Heidegger saw technology as responsible for 'the spiritual decline of the earth'. Nonetheless he was also the first philosopher to seriously think technology after Marx. As Mark Poster notices in his rereading of Heidegger's 1955 essay, 'The Question Concerning Technology'—a rereading that explicitly tests the validity of Heidegger's thought for an understanding of new technologies—Heidegger's antipathy towards technology was accompanied by an enormous sensitivity to the problem of technology itself. According to Poster, Heidegger was 'no simple technophobe.'[18] In fact, the central point of Heidegger's argument in 'The Question Concerning Technology' is not technology *per se* but modern humanity's way of being. Technology characterizes modern 'culture'—the term that Poster chooses for Heidegger's *Dasein*.[19] For Heidegger, humanity has to 'bring itself forth' in order to be, and it does so in part through its use of things—that is, through technology. In this context, technology is understood as a whole process of setting up the world. As long as humanity is aware of this process, it has a free relation to itself. This was the case in ancient Greece, where technology was openly visible and integrated into culture. However, in the modern age, technology has become a way of using things which brings humanity forth, while at the same time concealing this very process—that is, concealing technology itself. Ultimately, modern technology 'challenges' nature: it does violence to it and reduces it to an available resource. In so doing, it also reduces humanity to the same status, since humanity as part of nature becomes a servant of technology. Heidegger calls this process 'enframing'. His hope is that humankind recognizes this process of enframing and becomes capable of developing a kind of technology which would be completely different from today's.[20]

This brief, even schematic, synthesis of Heidegger's thought nevertheless allows us to see that he does not view technology as essentially instrumental. On the contrary, technology is for him a way of being in the world. This is the sense of his famous affirmation that 'the essence of technics is nothing technical.'[21] According to Stiegler, this is precisely what makes Heidegger's understanding of technology so interesting. Heidegger suggests that, if we keep thinking technology as a 'means', we will never be able to understand what technology *is*. In other words, we cannot think technology efficaciously as long as we remain in the frame of mind of instrumentality.

Heidegger's understanding of technology is deeply connected to his conception of time. For him, calculation has its roots in our relation to the future and in our attempt to determine future possibilities, which we fear precisely because they appear indeterminate. Heidegger describes this process as 'anticipation' or 'concern': our attempt to control (or to anticipate) the uncertainty of the future creates the basis for calculation, or for circumscribing the realm of possible futures. Understood in a broader historical context, this is what Heidegger identifies as the turning of Western thought into calculation in the modern age. This is also why for him technology has a central role in defining modernity. On the other hand, modern technology also opens up for us the possibility of radically reconceiving technology itself by becoming conscious of the instrumental approach which has characterized our understanding of technology since Aristotle. This is why Stiegler praises Heidegger as the first philosopher who dares to propose a radical reconceptualization of technology.[22]

Stiegler also identifies a 'Marxist offshoot' to Heidegger's thought—developed for instance by Jürgen Habermas—which nonetheless did not manage to escape an instrumental conception of technology.[23] It is worth mentioning Habermas's work briefly because his position seems to underlie much of the contemporary debate on the political and ethical aspects of new technologies. And yet, Habermas's theory of technology, albeit extremely interesting, does not account for the way in which contemporary technology works, and therefore does not contribute to its demystification as much as Heidegger's thought does. Unlike Heidegger, Habermas does not propose a radical reconceptualization of the philosophical conception of technology. He identifies in modern society a form of technocracy, that is, of political domination which is not recognizable as such because it derives its legitimation not from the political arena but rather from scientific and technical knowledge. More precisely, the nature of technocracy is to confuse political legitimation and techno-scientific legitimation, therefore making techno-scientific knowledge the only visible source of legitimacy, and transforming any opportunity for political debate into a malfunction in a society that needs to be fixed. Technocracy substantially depoliticizes society. Furthermore, language itself is 'technicized', since technical and scientific frames of mind spread all over society and include also communication. Habermas and Heidegger both conceptualize technical modernity as a paradoxical situation in which technology ends up doing a disservice to humanity rather than being in its service. Nevertheless, Habermas continues to analyze technology in terms of ends and means. He suggests that we pursue a liberation of language from its technicization, and that we turn technology into an object of democratic debate in a free language. On the contrary, Heidegger problematizes the very concept of 'means' and, much more radically, suggests that we rethink 'the bond originarily formed by, and between, humanity, technics, and language'.[24]

In sum, for Habermas, technology can be treated as an object of a discussion that takes place in a transparent language and that is based on what he calls 'good reasons'—that is, rationally convincing arguments. He seems to believe that technology does not have any real effect on language itself, or at the very least that the language of politics can be separated from the language of technoscience. Furthermore, he does not really take into account the transformation that information and communication technologies introduce in the 'public sphere', which is supposed to be the space of a democratic debate.[25] Even when examining biotechnologies and acknowledging their radical novelty, he does not seem to break free from the general framework of instrumentality.

For instance, in *The Future of Human Nature*, Habermas claims that biotechnologies generate unprecedented moral problems and that genetic technologies are capable of affecting what it means to be human.[26] Enhancement-oriented genetic practices seem to him to entail an asymmetrical relationship of influence between generations, since the genetic programming of a child threatens his future freedom when it comes to choosing and shaping his own destiny. The collective deliberative process that seems to be Habermas's favoured solution when dealing with both technology and politics does not function here because future generations cannot take part in it. Habermas's attempt to solve this problem consists in proposing the concept of 'species ethics'—that is, the domain of decisions made by the human species as a whole about the question of what it means to be human. Our concern for ensuring future persons' status as free and equal beings is situated by Habermas at the level of species ethics. Therefore, although he understands genetic technology as deeply troubling the very idea of humanity, he ultimately reinforces the latter by simply shifting the traditional values of individuality, equality, and freedom to the level of the species. What it means to be human remains unquestioned, and so do the possible political consequences of such questioning.

In contrast to Habermas, I want to reiterate my earlier point that a deeper understanding of technology and of its relation with the human is needed if we are to understand the political implications of new technologies. In other words, we need to rethink technology philosophically if we want to think it politically. The question is not one of discussing technology 'democratically' through a 'freed' language. It is rather a matter of recognizing the mutually constitutive implications of technology and language—via the concept of instrumentality—and therefore of radically rethinking both terms *together*, since there is no way of (re)thinking one without the other.

In order to clarify the concept of 'originary technicity' further and to investigate its significance for my analysis of new technologies, let me now return to Stiegler's work. What Clark calls 'originary technicity', Stiegler names 'originary prostheticity' of the human.[27] To clarify this concept further, it is helpful to examine the third volume of *Technics and Time*, particularly where—in dialogue with Derrida—Stiegler reworks Husserl's philosophy of time.[28] Stiegler's philosophy of technology is based on the central premise that the human has always been technological. Stiegler draws here on the work of the French paleontologist André Leroi-Gourhan, who tightly connects the appearance of the human with tool use. For Stiegler, too, the human coemerges with tool use. He writes:

> Human beings disappear; their histories remain. This is a huge difference from all other living beings. Among the various traces humans leave behind, some are products with entirely different ends from any 'conversation with memory': a clay pot, for example, is not a tool made to transmit memory. But it does so, spontaneously, nonetheless, which is why archaeologists consult it in their research: pots, etc., are often the only witnesses to the most ancient cultural *episodes*.[29]

From this perspective, technology carries the traces of past events. In Mark Hansen's words, it is 'the support for the inscription of memory'—that is, technology is always a memory aid, and only through memory do human beings gain access to their own past, and therefore become aware of themselves, or gain a consciousness.[30] Any technical instrument registers and transmits the memory of its use. For instance, a carved stone used as a knife preserves the act of cutting, thus becoming a support for memory. In this sense, technology is the condition of the constitution of our relation to the past.

In sum, it can be said that human beings 'exteriorize' their memory into technological objects, which in turn are nothing but memory exteriorized. Importantly, by doing this the human species becomes able to suspend its genetic program and to evolve through means other than animal instincts—that is, in Stiegler's words, to 'pursue life through means other than life'. Stiegler gives the name 'epiphylogenesis' to this process.[31] Epiphylogenesis is the transformation and evolution of the human species through its relationship with technology, rather than only on the basis of its genetic program. Furthermore, by functioning as a support for memory, a technical object for Stiegler forms the condition for the givenness of time in any concrete situation. For this reason, he maintains that human beings can experience themselves only through technology.[32] This formulation becomes much clearer if we consider cinema, which for Stiegler is the emblematic technology of contemporaneity. Hansen comments:

> More than any other technology (and certainly more than literature), it is cinema in its contemporary form as global television that frames time for us and gives us a surrogate temporal object in whose reflection we become privy to the flux of our own consciousness. At the same time, by opening consciousness onto the past, onto the non-lived tradition of historicality, onto otherness of that which does not belong to the experience of consciousness, cinema *qua* temporal object captures the contemporary manifestation of the interdependence of the who and the what, of the human subject and the technical other. Put bluntly, we become who we are by inheriting a past destined to us through cinema.[33]

For Stiegler, cinema (and, by extension, technology) makes available to us the experience of others, and therefore constitutes a striking example of the relation between technological objects and time. Hansen's complex passage, based on Stiegler's rereading of Husserl's phenomenological thought, which in turn constitutes the basis of Stiegler's analysis of the relation between technology and time, opens up a whole new series of questions in relation to my investigation of new technologies. First, by analogy, one could ask: what kind of temporality becomes accessible to us through new technologies, and more specifically through software? We could say that software-based technologies, such as real-time technologies, can, on the one hand, bypass the human perception of time.[34] On the other hand, different kinds of software-based technologies, such as word processors, operate on a much more 'human' scale of temporality. Moreover, common to all software-based technologies is the fact that, before they become operative, they must be programmed — that is, software must be designed and developed. Therefore, one could ask a second question: what kind of temporality do we access through programming and what kind of relation to ourselves do we establish through software? Here I want to suggest that Stiegler's rereading of Husserl's phenomenology can help with answering these questions.

The experience of others that we have not directly experienced but that becomes accessible to us because it has been recorded is what Stiegler calls 'tertiary memory'.[35] Stiegler draws here on Husserl's concept of 'image-consciousness'. An example of image-consciousness is for Husserl a painting 'where the artist . . . archives her experience in the form of a memory trace'.[36] This trace is an image of the past and of the memory of the artist, but it is not an image of the lived past of the viewer. While for this reason Husserl excludes image-consciousness from any role in time-consciousness, Stiegler reverses the argument: for him tertiary memory (that is, memory that has not been lived through by us) is the very condition of time-consciousness. In other words, Stiegler foregrounds a consequence of Husserl's thought that Husserl himself hesitated to recognize: namely, the intrinsically technical basis of our consciousness of time.

To be more specific, Husserl recognizes that we cannot grasp temporality by a direct analysis of consciousness, and that we necessarily need to examine 'an object that is itself temporal'. A temporal object is defined as 'an object that is not simply in time but is constituted through time and whose properly objective flux coincides with the flux of consciousness when it is experienced by a consciousness. Husserl's favoured example is a musical melody.'[37] As Hansen points out in his careful analysis of Stiegler's thought, by focusing on the temporal object Stiegler can complicate Husserl's analysis of time-consciousness and introduce technology into it at a deeper level than Husserl himself does — that is, at the level of 'primary retention'. What is important here is that Stiegler reverses

Husserl's hierarchy. We do not have a primary understanding of time and *then* technology: it is rather technology that gives us an understanding of time. The reason for this is that we always find ourselves in the midst of a horizon—a world already constituted by and comprising both what we had experienced in the past *and* the past we never experienced (but that was experienced by others and given to us through technical memory supports).[38] To recapitulate Stiegler's argument, the relationship between time and technology is for him a fundamental one, since technology constitutes the condition for our experience of time. But to what extent is this understanding of technology as recorded experience—that is, as memory—helpful in our investigation of software-based technologies and of software? If software is exteriorized memory, what does it record? Or—to rephrase the question—what would it mean to analyze software as a 'temporal object' (in Husserl's terms), or as 'a technical object' (in Stiegler's terms)?

The idea of the technical object was in fact formulated for the first time by the French palaeontologist André Leroi-Gourhan. In his 1964 book, *Le Geste et la Parole*, Leroi-Gourhan presents technology as a privileged point of access to the understanding of human evolution—indeed, as the pivot of a unified theory of human evolution. For him, anthropology must be founded on technology, understood as the study of human material culture. He identifies the vertical posture, the use of the hand for purposes other than locomotion, and the presence of tools and language as characteristics of the human species. For Leroi-Gourhan, freedom of the hand during locomotion implies the beginning of technical activity, in the same way that manual expertise frees the mouth from procuring nourishment and makes it available for speech. However simplified this synthesis, we can say that Leroi-Gourhan views evolution as a series of liberations taking place from the Paleozoic to the Quaternary eras, and from fish to human. Even more importantly, he asserts 'not only that language is a characteristic of humans as are tools, but also that both are the expression of the same intrinsically human property'.[39] Language and tools evolve together, for they are 'neurologically linked and cannot be dissociated within the social structures of humankind'.[40] In this context, Leroi-Gourhan proposes what is generally considered to be his fundamental contribution to anthropology and archaeology—that is, the concept of 'operating sequence' as a kind of sequential organization that underlies both language and technology. For Leroi-Gourhan, culture, that emerges with *Homo sapiens*, is also an operating sequence—a program, a process of exteriorization, the 'placing outside ourselves' (and consequent socialization) of what in the rest of the animal world remains at the level of instinct. Importantly, this process of exteriorization, both of tools and of memory, enables Leroi-Gourhan to integrate contemporary technology into a unitary process of biocultural evolution. For him, today evolution has reached the stage of the exteriorization of the brain in computers. He envisions a possible future in which human beings become simply outdated, and the future of the species *as species* resides in intelligent machines. Thus, not only does Leroi-Gourhan explain culture as the exteriorization of memory but ultimately he is able to integrate the transmission of exteriorized (or collective) memory—that is, of culture—into a historical progression through the stages of oral transmission, written transmission, and, finally, 'electronic transmission'. For him, computers belong to the latest phase, as the furthest example of the exteriorization of memory.

In sum, Leroi-Gourhan's theory proves rather interesting for an examination of software-based technologies, because it explicitly addresses the development of what I have called 'software-based technologies'—and what he calls 'intelligent machines'—from an anthropological point of view, and because it strives to position computers within the process of biocultural evolution.[41] In fact, he should be credited for stimulating the reconceptualization of software-based technologies within a general frame of reference that escapes the concept of instrumentality and that regards technology as constitutive of the human. However, it is important to notice that Leroi-Gourhan does not seem able to answer the very question he opens up—namely, the question of the emergence of the 'operational sequence' (and therefore of language, technology, and ultimately of humanity) as a result of the interaction between the brain and the physical environment. Therefore, as Stiegler shows in the first volume of *Technics and Time*, it is precisely as a general frame of reference—that is, as a theory of originary technicity—that Leroi-Gourhan's thought proves less satisfactory.[42]

In critical dialogue with Leroi-Gourhan, Stiegler proposes an alternative answer to the dilemma of the origin of technology and of the human. His solution is based on the concept of the 'technical object' as 'organized inorganic matter' (*matière inorganique organisée*)—a further clarification of the idea of technology as the support for consciousness. For Stiegler, organized inorganic matter is matter which transforms itself in time as technical object. These transformations are the condition of the human temporalization of time, in the sense that each time technological objects undergo radical evolution the temporalization of time changes. As Richard Beardsworth has convincingly pointed out, this concept of the technical object holds two orders of consequences: it allows us to think the transformation of technology through time, while at the same time exposing the crucial role that technology plays in the constitution of the human experience of time.[43] According to Stiegler, material objects from the stone instrument to the portable computer change our way of perceiving time, and therefore affect the emergence of our identity and our understanding of what it means to be human.

Thus, Stiegler's articulation of matter as 'inorganic organized matter' allows for a history of human culture as—in Beardsworth's terms—'the history of the differentiation of the originary complex human-technical.'[44] Beardsworth's formulation is significant because it shows exactly how Stiegler resolves Leroi-Gourhan's impasse on the origin of technology. For Stiegler, the mutual constitutivity of technology and the human makes it impossible to decide which is the origin of the other. Nevertheless, we can tell the history of how this reciprocal ongoing constitution takes place over time. We can investigate how human beings and technology coevolve without having to decide which one to prioritize. Stiegler names this process within which the human and the technical mutually constitute each other 'the originary complex who-what.'[45]

To summarize, human beings, technology, and culture are part of the same process of exteriorization, which, as we have seen earlier, Stiegler names 'ephiphylogenesis'. For him we have emerged as human beings as a 'result' of three kinds of memory: our genetic memory, the 'individual' memory—or the memory of our central nervous system (which is responsible for our remembering of experiences, and which Stiegler names 'epigenetic'), and the techno-logical ('epiphylogenetic') memory, which preserves the experiences of past generations in the tools and language we inherit from the past, and therefore is an 'externalized', shared memory.[46] With epiphylogenesis the human being reaches a new relationship with its environment—a relationship mediated through technology. Technology carries with it memories of the past—not only of the individual past, but of the past generations. In this sense, it can be said that the 'who' (the human being) invents technology, but at the same time it is invented ('instructed') by it, by the memory of past experiences that technology carries.[47] This is the sense of the mutual co-constitution of the 'who' and the 'what', of technology and the human. Stiegler's answer to Leroi-Gourhan's anthropological impasse on technology is to emphasize that there is actually no way to distinguish between the material aspects of technology and the temporality it carries with it—and therefore there is no way to separate technology from culture, or technology from the human. But what would it mean to investigate software as 'organized inorganic matter'? What would the consequences of thinking software within the framework of originary technicity be? In what way is software a 'what' that constitutes the 'who' that interacts with it? In what way is one constituted by software as much as one produces and uses it?

What I want to emphasize here is that, however important, Stiegler's reflections on originary technicity cannot be transferred into the analysis of software uncritically. The main question that needs to be addressed is the distinction operated by Stiegler between technics and mnemotechnics, which is fundamental to his understanding of contemporary technologies as a major break in the history of technology. In the third volume of *Technics and Time*, he explains:

> [The] independence of the *mnemo*technical relative to systems of technical production is no longer the case today: the new global technical system has become a *global mnemotechnical* system in which technical and mnemotechnical systems have fused and have become, at the same time, global. . . . The global technical system has become essentially a mnemotechnical system of industrial production of tertiary retentions, and thus of the retentional selection criteria for the flux of consciousness inscribed in the processes of adoption.[48]

For Stiegler, while all kinds of technology always transmit memories, some have been produced expressly with a view to transmitting memories. Stiegler gives the name of 'mnemotechnics' to technologies specifically devoted to the transmission of memory (for instance, writing, photography, phonography, and cinematography). He explains: '[T]echnics is before all else a memory support, what I have called epiphylogenesis. But not all technics is a *mnemo*technique: the first mnemotechnical systems appeared, it seems, after the Neolithic era, eventually to become the various forms of writing we know and use today.'[49]

In Stiegler's view, technical systems precede mnemotechnical systems. When he investigates the historical transformation of technology, he focuses on technical systems—that is, mainly technological systems of production. In the first volume of *Technics and Time*, he draws on Bertrand Gille's concept of 'technical system' as a moment of stability in time, or a point of equilibrium in the process of technical change that characterizes history.[50] This point of equilibrium is expressed in a particular technology. In other words, every civilization constitutes itself around a technical system, which is in turn organized around a dominant technology. Every technical system has in itself a potential for change, and actually undergoes evolutionary transformations and periods of crisis. During a crisis, a technical system evolves at great speed, causing 'dis-adjustments' with the other social systems—such as economy, politics, education, and so forth, and it can only return to (relative) stability when these other systems have 'adopted' the new technical system.

As we have seen above, for Stiegler the contemporary globalized industrial technical system (whose beginnings took root in England at the end of the eighteenth century) has entered an epoch of permanent innovation, becoming fundamentally unstable. Such globalization of the industrial technical system has been made possible, to a great extent, by information and communication technologies, which facilitate, for instance, the automation and control of remote production and distribution, the international circulation of capital, and the opening up of intercontinental markets. The contemporary global system represents a break in the history of technology precisely because of the newly acquired importance of information and communication technologies. Until recently, mnemotechnics had always evolved slower than the technical systems of material transformation. While the latter underwent substantial changes from the age of ancient Greeks to the Industrial Revolution, alphabetic writing remained more or less stable. This independence has now ceased to exist, since communication and information industries have become the centre of the technical system of production and—more generally—the decisive element of the global technological system. This has in turn led to a change in our perception of space and time. For instance, the distances and the delay in the circulation of messages tend to be nullified by global networks, and 'day and night themselves have become confused by the artificial light of the electric bulb and the cathode-ray tube.'[51] Our mechanisms of orientation are therefore profoundly disturbed.

The question of software as inorganic organized matter becomes thus the question of the place of software in the globalized mnemotechnical system. Such a question needs to be reformulated as follows: what kind of mnemotechnics is software? And—even more importantly—is Stiegler's distinction between technics and mnemotechnics meaningful for an investigation of software in the first place? To understand this extremely important point, one must be reminded that—as we have seen earlier on—while commenting upon the novel *The Turing Option*, Clark describes technology as 'systems of signs'; thus, for him, originary technicity seems to have an intrinsic relation to what Stiegler calls mnemotechnics.[52] Conversely, Stiegler's distinction between the two is based on the following argument: every technics (for instance, pottery) carries the memory of a past experience; but only mnemotechnics (for instance, writing) are conceived with the primary purpose of carrying the memory of a past experience. In Stiegler's argument, the emphasis is on the aim, or end, of different technologies: some technologies are conceived just for recording, others are not.

At this point I want to advance the following proposition: software transgresses Stiegler's distinction between technics and mnemotechnics. Although this thesis needs to be proved, it is important to position it first of all as a problem. Take, for instance, the definition of software as the totality of all computer programs as well as all the written texts related to computer programs given by software engineering.[53] According to this definition, software can be thought of as a totality of 'documents' or 'texts' written in natural and formal languages, and therefore—in Stiegler's terms—as mnemotechnics. On the other hand, it cannot be said that the main purpose of software is recording in the same way that it is for writing or cinema. It could be argued that the main purpose of software is to make things happen in the world (for instance, to change the polarities of the electronic circuits within a computer on which software is executed). This is why software might be the point where Stiegler's distinction between technics and mnemotechnics is suspended.[54]

To a certain extent, by introducing the distinction between technics and mnemotechnics, Stiegler involuntarily reintroduces the separation between the technical and the symbolic that he deconstructs in Leroi-Gourhan. Certainly, in the third volume of *Technics and Time*, Stiegler speaks of a convergence between technics and mnemotechnics, but in a much more general sense—that is, as a convergence of technologies of production with information and communication technologies. For him, undoubtedly, information and communication technologies fall under the rubric of mnemotechnics—or technology that has recording as its primary aim. Ultimately, in order to distinguish between technics and mnemotechnics, Stiegler resorts to the concept of the aim (or the end) of technology, therefore seemingly falling back into an instrumental con-

ception of technology—which obviously contradicts his understanding of technology as originary.

But, if software calls into question the distinction between technics and mnemotechnics, how is one supposed to think software
within the framework of originary technicity? In what way is the
relationship between the technical and the symbolic articulated in
software? As we have seen, Clark states that the thinkers of originary technicity situate the question of technology 'in the subtle intimacy' of the relation between technology and language. Moreover,
according to Clark, Jacques Derrida is one of the most important
thinkers of originary technicity precisely because he 'takes on the
radical consequences of conceiving technical objects (including systems of signs) as having a mode of being that resists being totally
understood in terms of some posited function or purpose for human being'.[55] By his refusal to explain either technology or language in instrumental, functionalist terms Derrida resists the widespread denigration of the 'merely' technical in Western thought. In
fact, Derrida makes references to technology and to the importance
of technicity for the definition of the human throughout his whole
work. Moreover, his conception of technology as something that
cannot be understood within the conceptual framework of instrumentality is inseparable from his understanding of writing. For
him, as for Stiegler, the devaluation of instrumentality that can be
traced back to Plato's *Phaedrus* cannot be separated from the devaluation of writing.

Famously, Derrida's reflection on writing is crucial to the whole
of his theory, and lies at the core of his criticism of Western metaphysics. Derrida's goal is not a reversal of priorities—namely, the
prioritizing of writing over speech—but a critique of the whole of
Western metaphysics that he understands as 'logocentric'. As Gaya-
tri Spivak points out in her introduction to *Of Grammatology*, the
term 'writing' is used by Derrida to name a whole strategy of investigation, not merely 'writing in the narrow sense' as a kind of notation on a material support.[56] Thus, Derrida writes *Of Grammatology*
not to pursue a mere valorization of writing over speech, but to
present the repression of writing 'in the narrow sense' as a symp-
tom of logocentrism that forbids us to recognize that everything is
pervaded by the structure of 'writing in general'—that is, an eternal
escaping of the 'thing itself'.[57] Derrida argues that speech too is
structured like writing. There is no structural distinction between
writing and speech—except that, in the history of metaphysics,
writing has been repressed and read as a surrogate of speech. In the

chapter 'The end of the book and the beginning of writing' of *Of Grammatology*, Derrida maintains that today writing can no longer be thought as 'a particular, derivative, auxiliary form of language in general', or as 'an exterior surface, the insubstantial double of a major signifier, *the signifier of the signifier.*'[58] Making writing instrumental is a move of Western metaphysics, and it is paired with the notion of speech as fully present. From this perspective, writing is seen as an interpretation of original speech, as technology in the service of language. However, Derrida suggests that language could only be a 'mode' or an aspect of writing.

Derrida's questioning of logocentrism goes hand in hand with his questioning of the instrumental conception of technology. In *Mémoires for Paul de Man*, he states that '[t]here is no deconstruction which does not . . . begin by calling again into question the dissociation between thought and technology, especially when it has a hierarchical vocation, however secret, subtle, sublime or denied it may be.'[59] Thus, once again, Derrida makes it explicit that the dissociation between thought and technology is—as is every other binary opposition—hierarchical, since it implies the devaluation of one of the two terms of the binary—in this case, technology. For this reason Clark suggests that 'originary technicity' can be considered another name for Derrida's 'writing in the general sense'.[60] As Derrida states in *Of Grammatology*: '[w]riting is not an auxiliary in the service of science—and possibly its object—but first, as Husserl in particular points out in *The Origin of Geometry*, the condition of the possibility of ideal objects and therefore of scientific objectivity. Before being its object, writing is the condition of the *episteme.*'[61] This passage is crucial for clarifying the relationship that Derrida establishes between writing and thought, and ultimately for his understanding of technology as constitutive of the human. As Clark explains, for Derrida 'writing enregisters the past in a way that produces a new relation to the present and the future, which may now be conceived within the horizon of an historical temporality, and as an element of ideality.'[62] Thus, the written mark gives us the possibility of keeping trace of the past and enables us to acquire a sense of time. Clearly Derrida views writing—understood here as technology, or the technological capacity of registering the past—as a constitutive condition of thought. Consequently, technology cannot be understood through the opposition between *technê* and *epistêmê*, because it precedes and enables such an opposition. But what would all this mean for an investigation of software? To be more specific, in what way would the reformulation of 'originary technicity' in terms of

Derrida's 'writing in general' advance our analysis of software? This reformulation of the problem amounts on the one hand—as we have already seen—to asking what the significance of the study of software for an understanding of the relationship between technology and the human is. On the other hand, it opens up the methodological question of whether and in what way software should be approached as a historically specific technology. In order to start addressing both of these questions, let me now examine Derrida's rereading of Leroi-Gourhan's work in *Of Grammatology*.

For Derrida, Leroi-Gourhan has shown in *Le geste et la parole* that the historical perspective that associates humanity with the emergence of writing (and therefore excludes peoples 'without writing' from history) is profoundly ethnocentric. In fact, it shortsightedly denies the characteristic of humanity to peoples that do not actually lack 'writing', but only 'a certain type of writing'—that is, alphabetic writing.[63] To explain this point Derrida draws on Leroi-Gourhan's concept of 'linearization'. For Leroi-Gourhan, the emergence of alphabetic writing must be understood as a process of linearization. In his analysis of the emergence of graphism, Leroi-Gourhan emphasizes what he considers to be the underestimated link between figurative art and writing. '[I]n its origins', he states, 'figurative art was directly linked with language and was much closer to writing (in the broadest sense) than to what we understand by a work of art.'[64]

Given the difficulty of separating primitive figurative art from language, he proposes the name 'picto-ideography' for this general figurative mindframe. Yet he is very clear that such a mindframe does not correspond to writing 'in its infancy.'[65] Such an interpretation would amount to applying to the study of graphism a mentality influenced by four thousand years of alphabetic writing—something that linguists have often done, for instance, when studying pictograms. But 'picto-ideography' signals an originary independence of graphism from the mental attitude that constitutes the basis of what Leroi-Gourhan calls 'linearization'.

To understand the concept of linearization better, one must start from Leroi-Gourhan's concept of language as a 'world of symbols' that 'parallels the real world and provides us with our means of coming to grips with reality'.[66] For Leroi-Gourhan, graphism is not dependent on spoken language, although the two belong to the same realm. Leroi-Gourhan views the emergence of alphabetic writing as associated with the technoeconomic development of the Mediterranean and European group of civilizations. At a certain point in time during this process, writing became subordinated to spoken language. Before that—Leroi-Gourhan states—the hand had its own language, which was sight-related, while the face possessed another one, which was related to hearing. He explains:

> At the linear graphism stage that characterizes writing, the relationship between the two fields undergoes yet another development. Written language, phoneticized and linear in space, becomes completely subordinated to spoken language, which is phonetic and linear in time. The dualism between graphic and verbal disappears, and the whole of human linguistic apparatus becomes a single instrument for expressing and preserving thought—which itself is channelled increasingly toward reasoning.[67]

By becoming a means for the phonetic recording of speech, writing becomes a technology. It is actually placed at the level of the tool, or of 'technology' in its instrumental sense. As a tool, its efficiency becomes proportional to what Leroi-Gourhan views as a 'constriction' of its figurative force, pursued precisely through an increasing linearization of symbols. Leroi-Gourhan calls this process 'the adoption of a regimented form of writing' that opens the way 'to the unrestrained development of a technical utilitarianism'.[68]

Expanding on Leroi-Gourhan's view of phonetic writing as 'rooted in a past of nonlinear writing', and on the concept of the linearization of writing as the victory of 'the irreversible temporality of sound', Derrida relates the emergence of phonetic writing to a linear understanding of time and history.[69] For him, linearization is nothing but the constitution of the 'line' as a norm, a model—and yet, one must keep in mind that the line is *just* a model, however privileged. The linear conception of writing implies a linear conception of time—that is, a conception of time as homogeneous and involved in a continuous movement, be it straight or circular. Derrida draws on Heidegger's argument that this conception of time characterizes all ontology from Aristotle to Hegel—that is, all Western thought. Therefore, and this is the main point of Derrida's thesis, 'the meditation upon writing and the deconstruction of the history of philosophy become inseparable.'[70]

However simplified, this reconstruction of Derrida's argument demonstrates how, in his rereading of Leroi-Gourhan's theory, Derrida understands the relationship of the human with writing and with technology as constitutive of the human rather than instru-

mental. Writing has become what it is through a process of linearization, and in doing so it has become instrumental to speech. Since
the model of the line also characterizes the idea of time in Western
thought, questioning the idea of language as linear implies questioning the role of the line as a model, and thus the concept of time
as modeled on the line. It also implies questioning the foundations
of Western thought (by means of a strategy of investigation that, as
we have seen, Derrida names 'writing in general', or 'writing in the
broader sense'). At this point it becomes clear why, if we follow
Derrida's approach to originary technicity, a new understanding of
technology (as intimately related to language and writing) entails a
rethinking of Western philosophy—ambitious as this task may be.

It is worth noting here that in *Of Grammatology* Derrida expressly
highlights how the reconceptualization of the Western tradition of
thought is particularly urgent today. Such a rethinking is what Derrida famously calls 'the end of the book', or the end of linear writing. According to Derrida, we are suspended today between two
eras of writing—and this is why we can also reread our past differently. Actually, the 'uneasiness' of philosophy in the past century is
due to an increasing destabilization of the model of the line. He
states that what is thought today cannot be written in a book—that
is, it cannot be thought through with a linear model—any more
than contemporary mathematics can be taught with an abacus. This
inadequacy does not only apply to the current moment in time, but
it comes to the fore today more clearly than ever. Derrida writes:

> The history of writing is erected [by Leroi-Gourhan] on the base
> of the history of the *grammé* as an adventure of relationships
> between the face and the hand. Here, by a precaution whose
> schema we must constantly repeat, let us specify that the history
> of writing is not explained by what we believe we know of the
> face and the hand, of the glance, of the spoken word, and of the
> gesture. We must, on the contrary, disturb this familiar knowl-
> edge, and awaken a meaning of hand and face in terms of that
> history.[71]

For Derrida, what is most relevant in Leroi-Gourhan's history of
writing is that it problematizes our conception of the human ('what
we believe we know of the face and the hand'). Yet the focus of
Derrida's work is not the concrete analysis of historical systems of
writing, since, as we have seen, he differentiates 'writing in general'
from any such system. With regard to my investigation of software,
then, Derrida's understanding of what Clark calls 'originary tech-

nicity' has two important implications. On the one hand, it confirms the fundamental relationship between technology and the human, and it supports the need for a radical questioning of both concepts—and ultimately of Western thought. On the other hand, Derrida leaves open the question of how to investigate a historically specific technology (for instance, software) without losing its significance for a radical rethinking of the relationship between technology and the human. It is actually Stiegler's fundamental rereading of Derrida's thought in *Technics and Time* that allows for such an investigation. In order to understand this important point further, it is now worth returning once again to the examination of Stiegler's work.

For Stiegler, the emergence of the technique of linear writing radically transforms the modes of cultural transmission from generation to generation. In fact, from the point of view of Greek pre-Socratic thought, which does not presume the immortality of the soul, the dead can nevertheless return as ghosts that transmit an inheritance, and such inheritance is deemed to come from a spirit (*esprit*) that crosses generations. This is the pre-Socratic image of cultural transmission. In contrast, the appearance of linear writing allows for the transmission of culture 'as a unified spirit, precisely through the unification of language enabled by literalization'.[72] Drawing once again on Leroi-Gourhan and Derrida, Stiegler insists that the emergence of the model of the line has changed both the transmission of culture *and* the modes of thought.

According to Stiegler, the Sophists themselves are a by-product of this process. The years between the seventh and the fifth century BCE are witness to the arrival of the *grammatists*, the masters of letters, and later on, of the Sophists, who 'go on systematically to develop a technique of language that quickly acquires a critical dimension, in so far as this technique of developed language will in turn engender a moral crisis'.[73] Thus, sophistry is not an oral technique; rather, it presupposes writing.[74] Accordingly, Plato criticizes the Sophists because they manage to speak well, 'but they learn everything by heart, by means of this techno-logical "hypomnesis" that is logography, the preliminary writing out of speeches. It is because writing exists that the Sophists can learn the apparently "oral" technique of language that is rhetorical construction.'[75] In the *Ion* Plato even makes a connection between poets and Sophists, claiming that they work along the same lines of falsehood: '[s]ophists, poets, are only liars, *that is to say*, technicians.'[76] This powerful image of the technician as a liar constitutes the summation of Plato's devaluation of technology and writing.

To summarize, Stiegler points out that, on the one hand, the question of technology, considered as the object of repression, 'is a question that emerges *with* and *by* its denunciation by Plato'. It arises 'above all as a *denial*, and in this sense therefore as *a kind of forgetting*' — and this is quite paradoxical, since in *Phaedrus* what Plato blames technology for is precisely its power of forgetting.[77] On the other hand, it can be said that the question of technology appears well before Plato: as we have just seen, it arises in the context of the transformation of the Greek cities, associated with the development of navigation, money, and above all mnemotechnics, that is to say of technologies capable of transforming the conditions of social and political life and of thought. Ultimately, *technê* and *epistêmê* — that is, knowledge and technology — share a relationship with writing, the fundamental mnemotechnics. In turn, mnemotechnics, and technology in general, reveal a constitutive connection with temporality.

Stiegler's understanding of the transformation of technology in time is crucially related to his 'displacement' of deconstruction that also results in his break with Heidegger. He explains:

> Let's say, for example, that one night I write the sentence: 'it is dark'. I then reread this sentence twelve hours later and I say to myself: hang on, it's not dark, it's light. I have entered into the dialectic. What is to be done here? . . . That which makes consciousness be self-consciousness (i.e. consciousness that is conscious of contradiction with itself) is the fact that consciousness is capable of externalising itself.[78]

This passage is extremely important because it reformulates the concept of the technical constitution of consciousness that Clark explores in his analysis of *The Turing Option*.[79] Here what Stiegler — and Leroi-Gourhan before him — calls 'exteriorization' (which constitutes the basis of self-consciousness) is clearly pursued *through writing*. One writes 'it is dark', and when one rereads the note twelve hours later it is light. This produces, as Stiegler himself further clarifies, 'a contradiction between times', namely the time of consciousness when one wrote this and the time of consciousness when one reads this. Yet, one has the same consciousness, which is therefore 'put in crisis', and this crisis in turn raises self-awareness. The act of inscription — that is, of exteriorization — ultimately constitutes interiority, which does not precede exteriority, and vice versa.[80] As I have explained earlier, for Stiegler (again drawing on Leroi-Gourhan) the process of exteriorization constitutes the foundation of temporality, of language and of technical production, and requires a basic neurological 'competence'—that is, 'a level of suitable cortical and subcortical organization'.[81] This is Stiegler's fundamental point of departure from Derrida's thought. Through this departure he lays the foundation for the concrete study of historically specific technologies as fundamental to the understanding of the constitutive relationship between technology and the human.

Stiegler's interpretation of the myth of Prometheus and Epimetheus in the first volume of *Technics and Time* clarifies this point even further. According to the myth, Zeus gives Prometheus the task of distributing qualities and powers to the living creatures, but Prometheus leaves it to his twin brother Epimetheus to act in his place. Epimetheus hands out all the qualities to the living creatures and forgets to keep one for the human being. Human beings therefore appear here as characterized by a 'lack of quality'. Stiegler comments that the human being is 'a being by default, a being marked by its own original flaw or lack, that is to say afflicted with an original handicap'.[82] For this reason, Prometheus decides to steal technology—that is, fire—and gives it to human beings, in order to enable them to invent artefacts and to become capable of developing all qualities. With the gift of technology, a problem arises: mortals cannot agree on how to use artefacts, and consequently start fighting and destroying each other. In Stiegler's words, '[t]hey are put in charge of their own fate, but nothing tells them what this fate is, because the lack [*défaut*] of origin is also a lack of purpose or end.'[83] Stiegler's reworking of the myth clearly shows how for him technology raises the problem of decision, and how this encounter of the human with decision in turn constitutes what he calls 'technical time'. Technical time emerges because human beings experience their capacity of making a difference in time through decision. Temporality is precisely this opening of the possibility of a decision, which is also the possibility of giving rise to the unpredictable, the new.

It is for this very reason that the historical specificity of technology is central to Stiegler's thought. The human capability of deciding 'what to become' constitutes temporality. Moreover, human prostheticity—that is, the fact that human beings, to survive, require nonliving organs such as houses, clothes, sharpened flints, and all that Stiegler calls 'organized inorganic matter'—forms the basis for memory, or better, for technical memory. As I have shown earlier

on, unlike genetic and individual memory, technical memory coincides with the process of exteriorization that 'enables the transmission of the individual experience of people from generation to generation, something inconceivable in animality'.[84] This inherited experience is what Stiegler calls 'the world'—that is, a world that is
always already haunted by 'spirits' in the pre-Socratic sense, always
already constructed by the memories of others.

Stiegler departs from Heidegger precisely in his understanding
of temporality. To simplify Stiegler's complex argument, his disagreement with Heidegger revolves around the different importance that the two philosophers give to the historical specificity of
technology. For Heidegger, temporality is originally technical, since
to be a temporal being—that is, to exist—one has to be in the world,
which for Heidegger is fundamentally the world of tools—or of
technology. Nevertheless, Heidegger believes that the most authentic temporality is experienced by human beings as a relation to
death. As human beings, we are structurally ignorant of the time
and place of our own death, and this relation to death plunges us
into an 'absolute indetermination'.[85] We do not know the end of our
life, both in the sense of its limit and of its meaning. In Stiegler's
terms, the content of our life is determined only after our death, that
is, 'too late', when we are not able to witness it. According to Heidegger, human beings try to flee the permanent anguish of the inde-
terminacy of their death, and ultimately of their future. This is what
Heidegger names 'concern'—namely, the human beings' attempt to
foresee their unforeseeable future, to make certain the uncertain, to
calculate something that is not calculable. Technology is part of this
process, precisely because it is a means of controlling the future.
Every technical field, from weather forecasts to financial analysis,
attempts such calculation. But any such attempt tends to obscure
human beings' relation to death, and for this reason Stiegler ultimately finds Heidegger's argument inconsistent—since, on the one
hand, Heidegger views temporality as originally technical, while on
the other hand, he believes that technology obscures 'authentic'
temporality.

In sum, Stiegler's departure from Heidegger, as well as his break
with Derrida, are based on Stiegler's own attention to the historic
specificities of technology. He pays close attention to the fact that
human beings, as beings who deal with decision, 'are continuously
called into question by the development of technics which overtake
them'.[86] Here Stiegler brings to the fore the problem of making
decisions regarding technology with which I have opened *Software*

Theory—that is, the problem of how to think technology in a politically meaningful way. The term 'overtaking' is deployed by Stiegler with reference, once again, to Leroi-Gourhan's and Gille's thought, as well as Simondon's: 'people form technical objects, he argues, but these objects, because they themselves form a dynamic system, go on to overtake their makers.'[87] Technical objects form a 'system' because none of them is ever thinkable in isolation: a cassette— Stiegler exemplifies—is of no use without a tape recorder, which in turn is of no use without a microphone, electricity, and so on. Such systems are also dynamic: they change according to different tendencies that combine within society, and are negotiated through processes of decision.

I want to highlight here how Stiegler's approach is extremely helpful in order to contextualize the necessity of making decisions about technology in the broadest possible perspective. In fact, such decisions do not just affect technology; they also change our experience of time, our modes of thought and, ultimately, our understanding of what it means to be human. On the one hand, if understood as originary, technology constitutes our sense of time—or, even better, we only gain a sense of time and memory, and therefore of who we are, through technology. On the other hand, technology as a system tends to overwhelm us, making it difficult to make decisions. Ultimately, we gain a sense of time through technology, and in turn every change in technology changes our sense of time— and this then changes the meaning that we give to the fact of being human. For this reason thinking technology in a politically effective and meaningful way involves much more than, for instance, discussing technology in a neutral and 'free' language, as Habermas would have it. Rather, it implies a radical problematization of the meaning of humanity. In fact, by making decisions on technology, we make decisions about what it means to be human. Ultimately, this is the sense of my affirmation that it is necessary to think technology *philosophically* in order to think it *politically*.

To conclude, Stiegler's rereading of Derrida calls for a concrete analysis of historically specific technologies while keeping open the significance of such an analysis for a radical rethinking of the relationship between technology and the human. It therefore enables us to regard software in its historical specificity without losing the possibility of investigating its relationship to 'originary technicity' and, ultimately, to the question of what it means to be human. Whether and in what way such an investigation can be pursued will be the focus of the next chapter.

NOTES

1. Stuart Hall, "Cultural Studies and Its Theoretical Legacies," in *Cultural Studies*, ed. Lawrence Grossberg et al. (New York and London: Routledge, 1992), 284.

2. S. Hall, "Cultural Studies and Its Theoretical Legacies", 278.

3. Gary Hall, *Culture in Bits: The Monstrous Future of Theory* (London and New York: Continuum, 2002). Hall expands on this point in subsequent works, particularly G. Hall and Clare Birchall, eds., *New Cultural Studies: Adventures in Theory* (Edinburgh: Edinburgh University Press, 2006), and G. Hall, *Digitize This Book! The Politics of New Media, or Why We Need Open Access Now* (Minneapolis: University of Minnesota Press, 2008).

4. Bernard Stiegler, *Technics and Time, 1: The Fault of Epimetheus* (Stanford, CA: Stanford University Press, 1998), ix.

5. Bernard Stiegler, *Technics and Time, 1*, 2.

6. Stiegler, *Technics and Time, 1*, xi.

7. Richard Parry, *The Stanford Encyclopedia of Philosophy*, s.v. "Episteme and Techne," 2003, http://plato.stanford.edu/archives/sum2003/entries/episteme-techne/.

8. See *Nicomachean Ethics* 6, 3–4 (Aristotle, *The Complete Works* [Princeton, NJ: Princeton University Press, 1968]).

9. Timothy Clark, "Deconstruction and Technology," in *Deconstructions. A User's Guide*, ed. Nicholas Royle (Basingstoke: Palgrave, 2000), 238.

10. During the ascent of Nazism in Germany, Husserl conceptualized the emergence of algebra (which had been ongoing since Galileo's times) as a technique of calculation that emptied geometry of its visual content. According to Husserl, by becoming viable to calculation, geometry renounces its capacity of visualizing geometrical shapes—or, in Husserl's terms, 'spatio-temporal idealities' (Edmund Husserl, *The Crisis of European Sciences and Transcendental Phenomenology. An Introduction to Phenomenological Philosophy* [Evanston: Northwestern University Press, 1970], 44–45). Therefore, as Stiegler comments, 'the technicization of science constitutes its eidetic blinding' (Stiegler, *Technics and Time, 1*, 3). What must be emphasized here is that 'calculation' is a constitutive part of the concept of instrumentality.

11. Stiegler, *Technics and Time, 1*, 3.

12. Plato, *The Collected Dialogues* (Princeton, NJ: Princeton University Press, 1989), 275–77.

13. Clark, "Deconstruction and Technology," 238; Richard Beardsworth, *Derrida and the Political* (New York: Routledge, 1996), 157.

14. Harry Harrison and Marvin Minsky, *The Turing Option* (London: ROC, 1992).

15. Clark, "Deconstruction and Technology," 240.

16. Clark, "Deconstruction and Technology," 240.

17. Stiegler, *Technics and Time, 1*, 4.

18. Mark Poster, "High-tech Frankenstein, or Heidegger Meets Stelarc," in *The Cyborg Experiments: The Extensions of the Body in the Media Age*, ed. Joanna Zylinska (London and New York: Continuum, 2002), 17.

19. Poster, "High-tech Frankenstein, or Heidegger Meets Stelarc," 18.

20. Poster, "High-tech Frankenstein, or Heidegger Meets Stelarc," 18.

21. Martin Heidegger, *The Question Concerning Technology and Other Essays* (New York: Harper and Row, 1977), 35.

22. Stiegler *Technics and Time, 1*, 7.

23. Stiegler *Technics and Time, 1*, 10.

24. Stiegler, *Technics and Time, 1*, 13.

25. Jürgen Habermas, *The Theory of Communicative Action* (Cambridge: Polity Press, 1991). See also Habermas, *The Postnational Constellation* (Cambridge: Polity Press, 2001).

26. Jürgen Habermas, *The Future of Human Nature* (Cambridge: Polity Press, 2003).

27. Clark, "Deconstruction and Technology," 240; Stiegler, *Technics and Time, 1*, 98–100.

28. Bernard Stiegler, *Technics and Time, 3: Cinematic Time and the Question of Malaise* (Stanford, CA: Stanford University Press, 2011).

29. Stiegler, *Technics and Time, 3*, 131.

30. Mark B. N. Hansen, "'Realtime Synthesis' and the Différance of the Body: Technocultural Studies in the Wake of Deconstruction," *Culture Machine* 5 (2003), n.p., http://www.culturemachine.net/index.php/cm/article/view/9/8.

31. Stiegler, *Technics and Time, 1*, 17.

32. Such an experience of the self is what philosophers have called 'self-affection' (Kant) or—and this is particularly important in Stiegler's thought—'internal time-consciousness' (Husserl).

33. Hansen, "'Realtime Synthesis' and the Différance of the Body."

34. 'Real-time' technologies are technically defined as 'technologies for which time is critical', and they are typically involved in controlling complex apparatuses in potentially dangerous situations, such as airplanes during flight. Since these systems respond to changes in their environments as soon as they happen, sometimes at a speed of milliseconds, the time of response belongs to an order of temporality practically unperceivable by human beings. Paul Virilio has written extensively on the inhuman temporality of contemporary technologies. See, for instance, Paul Virilio, *The Vision Machine* (Bloomington: Indiana University Press, 1994) and *Open Sky* (London: Verso, 1997).

35. Mark B. N. Hansen, *New Philosophy for New Media* (Cambridge, MA: MIT Press, 2004), 254.

36. Hansen, *New Philosophy for New Media*, 316.

37. Hansen, *New Philosophy for New Media*, 254.

38. In Mark Hansen's words, for Stiegler 'tertiary memories—meaning, basically, all experience that has been recorded and is reproducible—represent our means of inheriting the past, the prosthetic already-there, and, for this reason, actually condition the other two forms of memory. Stiegler emphasizes the technical specificity of tertiary memory, for it is only once consciousness has the capacity to experience the exact same recorded experience more than once that we can appreciate how secondary retention (the memory of the first or earlier experience[s]) influences a subsequent primary retention' (Hansen, "'Realtime Synthesis' and the Différance of the Body"). Stiegler's complex argument, which mobilizes and transforms Husserl's theory of time-consciousness, is an extension of Derrida's own deconstruction of Husserl's distinction between primary retention and recollection (or secondary retention), and therefore between perception and imagination, and between presence and absence.

39. André Leroi-Gourhan, *Gesture and Speech* (Cambridge, MA: MIT Press, 1993), 113.

40. Leroi-Gourhan, *Gesture and Speech*, 114.

41. Interestingly, Leroi-Gourhan mentions the principle of the Jacquard loom as the model for early automatic machines based on punched cards, and ultimately for computers. For him, computers are nothing but books that have become progressively autonomous from their human reader by becoming ca-

pable of sorting out documentary material according to the same principles that inform loom weaving. The role of the loom in the history of the computer is well known. (See, for instance, Philip Morrison and Emily Morrison, eds., *Charles Babbage and His Calculating Engines: Selected Writings by Charles Babbage and Others* [New York: Dover, 1961].) Sadie Plant writes that '[t]he loom is the vanguard site of software development' (Sadie Plant, "The Future Looms: Weaving Women and Cybernetics," in *Cyberspace/Cyberbodies/Cyberpunk: Cultures of Technological Embodiment*, ed. Mike Featherstone and Roger Burrows [London: Sage, 1995], 46; see also Plant, *Zeros and Ones: Digital Women and the New Technoculture* [London: Fourth Estate, 1998]). Fernand Braudel describes the loom as 'the most complex human engine of them all' (Fernand Braudel *Capitalism and Material Life 1400–1800* [London: Weidenfeld and Nicolson, 1973], 247). Jacquard's automated loom was based on the principle of punched cards, where the threads selected by each card were the ones that passed through its holes. This principle was not new (it had been used since early eighteenth century), but Jacquard strung the cards together in sequences, each of which constituted an ordered set of weaving operations—or, in Leroi-Gourhan's words, a program. For this reason, when in the 1840s Charles Babbage started working on his Analytical Engine (which is generally considered the prototype of the computer), he modelled it on Jacquard's strings of punched cards. His own contribution was to introduce 'the possibility of bringing any particular card or set of cards onto use *any number of times successively in the solution of one problem'* (Morrison and Morrison, *Charles Babbage and His Calculating Engines*, 264; original emphasis). Thus, Babbage considered his machine as characterized by memory and foresight—that is, by the possibility of referring to past operations in order to act in the future. However limited, this account of the relationship between early computer prototypes and the principles of the loom shows the presence of an explicit connection between memory and time even in the early days of computing.

42. Although Stiegler's understanding of technology as 'exteriorization' parallels Leroi-Gourhan's paleontological theory, he clearly demonstrates that Leroi-Gourhan's account of the coevolution of the human and the technological falls prey to the same difficulty that, in the eighteenth century, Jean-Jacques Rousseau confronted in 'The Origin of Languages'—namely, the classical question of 'the origin of man' (Stiegler, *Technics and Time, 1*, 117–41). To simplify Stiegler's argument, both Rousseau and Leroi-Gourhan resort to a providentialist explanation of the origins of humanity (Rousseau through the intervention of God, Leroi-Gourhan through the emergence of the human capability of using symbols).

43. Stiegler, *Technics and Time, 1*, 49. Richard Beardsworth, "From a Genealogy of Matter to a Politics of Memory: Stiegler's Thinking of Technics," *Tekhnema: Journal of Philosophy and Technology* 2 (1995): 87.

44. Beardsworth, "From a Genealogy of Matter to a Politics of Memory," 88.

45. Stiegler, *Technics and Time, 1*, 141.

46. Stiegler, *Technics and Time, 1*, 185.

47. Stiegler, *Technics and Time, 1*, 185.

48. Stiegler, *Technics and Time, 3*, 133.

49. Stiegler, *Technics and Time, 3*, 131.

50. Stiegler, *Technics and Time, 1*, 30–31.

51. Stiegler, *Technics and Time 3*, 134.

52. Clark, "Deconstruction and Technology," 240.

53. Ian Sommerville, *Software Engineering* (Boston: Addison-Wesley, 2011), 4.

54. Stiegler's recent substitution of the distinction between 'mnemotechnics' and 'mnemotechnologies' for the one between 'technics' and 'mnemotechnics'

serves mainly to emphasize the fact that technology is always a support for memory, and therefore it does not solve the problem. See, for instance, Bernard Stiegler, "Memory," in *Critical Terms for Media Studies*, ed. Mark B. N. Hansen and W. J. T. Mitchell (Chicago and London: University of Chicago Press, 2010), 70.

55. Clark, "Deconstruction and Technology," 240.

56. Jacques Derrida, *Of Grammatology* (Baltimore: The Johns Hopkins University Press, 1976), lxix.

57. On the other hand, in the section of *Of Grammatology* about Lévi-Strauss, Derrida suggests that no definite distinction between writing in the 'narrow' and the 'general' sense can be traced, for one slips into the other.

58. Derrida, *Of Grammatology*, 7.

59. Jacqued Derrida, *Mémoires for Paul de Man* (New York: Columbia University Press, 1986), 108.

60. Clark, "Deconstruction and Technology," 241.

61. Derrida, *Of Grammatology*, 27.

62. Clark, "Deconstruction and Technology," 241.

63. Derrida, *Of Grammatology*, 83.

64. Leroi-Gourhan, *Gesture and Speech*, 190.

65. Leroi-Gourhan, *Gesture and Speech*, 195.

66. Leroi-Gourhan, *Gesture and Speech*, 195.

67. Leroi-Gourhan, *Gesture and Speech*, 210.

68. Leroi-Gourhan, *Gesture and Speech*, 212.

69. Derrida, *Of Grammatology*, 85.

70. Derrida, *Of Grammatology*, 86.

71. Derrida, *Of Grammatology*, 84.

72. Bernard Stiegler, "Technics of Decision: An Interview," *Angelaki* 8, no. 2 (2003), 154.

73. Stiegler, "Technics of Decision," 155.

74. Stiegler's assertion mirrors Derrida's argument that we need to have a sense of writing in order to have a sense of orality. I will return to this point in Chapter 2.

75. Stiegler, "Technics of Decision," 155.

76. Stiegler, "Technics of Decision," 155.

77. Stiegler, "Technics of Decision," 155.

78. Stiegler, "Technics of Decision," 163.

79. Clark, "Deconstruction and Technology," 240.

80. Stiegler, "Technics of Decision," 163.

81. Stiegler, "Technics of Decision," 164.

82. Stiegler, "Technics of Decision," 156.

83. Stiegler, "Technics of Decision," 156.

84. Stiegler, "Technics of Decision," 159.

85. Stiegler, "Technics of Decision," 159.

86. Stiegler, "Technics of Decision," 162.

87. Stiegler, "Technics of Decision," 162.

TWO

Language, Writing, and Code

Towards a Deconstructive Reading of Software

In his study of 2000 on 'Deconstruction and Technology', Timothy Clark observes:

> Derrida's arguments on technology affirm an originary technicity or supplementarity both constituting the human and transgressing any would-be pure or essentialist distinctions between concepts of the human and the machine. At the same time . . . [deconstruction] works to reveal and undo the profound complicity between metaphysical humanism and projects which idealize the technical as the correlate of a totally assured system of formal elements whose syntax or mechanics can be calculated — the notion of a technics which can be completely subordinated to logic.[1]

In this complex passage, Clark emphasizes how Jacques Derrida — who, as we have seen in Chapter 1, can be considered one of the most important thinkers of originary technicity because he refuses to explain either technology or language in instrumental (functionalist) terms, thus resisting the widespread denigration of the 'merely' technical in Western thought — pursues a double strategy when examining technology. On the one hand, he recognizes the foundational (originary) character of technology for the human, while on the other hand, he reveals technology's complicity with metaphysics. I want to take Derrida's approach as the point of departure for my argument in this chapter, which aims at establishing whether and in what way software can be studied as a historically specific technology without overlooking its relevance for an understanding of originary technicity.

The first aspect of Derrida's double strategy consists in the capacity of deconstruction to upset 'received concepts of the human and the technological by affirming their mutual constitutive relation or, paradoxically, their constitutive disjunction'.[2] Neither term can be said to take precedence over the other. Thus, technology cannot be understood as a tool for the human; nor can the human be understood as an effect of technology. 'The identity of humanity', Clark specifies, 'is a differential relation between the human and technics, supplements and prostheses.'[3] The second aspect of Derrida's strategy is his questioning of technology and technicist thought, in order to reveal their complicity with metaphysical humanism, and particularly with the instrumental conception of technology that characterizes Western philosophy. In Derrida's words:

> Computer technology, data banks, artificial intelligences, translating machines, and so forth, all these are constructed on the basis of that instrumental determination of a calculable language. Information does not inform merely by delivering an information content, it gives form, 'in-formiert', 'formiert zugleich'. It installs man in a form that permits him to ensure his mastery on earth and beyond.[4]

In fact, the two parts of Derrida's strategy cannot be easily separated. It is precisely by unmasking and undoing—or deconstructing—the complicity between metaphysical humanism and the 'idealization' of technology as something totally predictable ('calculable') that one can make originary technicity—that is, the constitutive relation between technology and the human—apparent. What I want to emphasize here is that Derrida's double strategy can help me in reformulating the question of how to study software as such: whether and in what way does software display a certain complicity with the instrumental understanding of technology? This question implies a second one—namely, to what extent, by undoing this complicity, light can be shed on the relationship between software (and more broadly technology) and the constitution of the human. Although these questions will find a definitive answer only in the following chapters, in order to set out the terms of the problem it is worth examining here at length Derrida's insight on new technologies and his clarification of how difficult it is to conceive them merely in terms of instrumentality.

Quite strikingly, for Derrida new technologies are already 'in deconstruction' in what he terms today's 'technological condition'. In his conversation with Bernard Stiegler in *Echographies of Television*, Derrida awards this process of deconstruction quite a broad meaning. For him, the acceleration of technological innovation in the contemporary world, coupled with the development of the technologies of the global media system (which Derrida and Stiegler group under the rubric of 'tele-technologies'), constitute a 'practical deconstruction' of the traditional political concepts of the public, the state, the citizen, and ultimately of the instrumental conception of technology itself.[5] On the one hand, in the contemporary world, technological innovation is massively appropriated by multinational corporations and nation-states and constantly programmed to support an economy of continual obsolescence. On the other hand, although programmed and neutralized as controlled 'development', technological innovation still gives rise to unforeseen effects. Derrida even propounds that the greater the attempt to control innovations, the more unforeseeable the future becomes. This second point is well exemplified for him by the relationships between telecommunications and the transformation of the public space in the late 1980s and early 1990s in Eastern Europe.[6] However, Derrida does not see this process as the realization of liberal humanist values, but as 'an accelerating process of spectralization'[7] — that is, as he explains in *Specters of Marx*, as 'the new speed of *apparition* . . . of the simulacrum, the synthetic or prosthetic image, and the virtual event, cyberspace and surveillance, the control, appropriations, and speculations that today deploy unheard-of powers'.[8] New media have 'spectral effects': from the dissemination of images and information beyond the borders of territorially delimited communities to the dependence of politics on its models of mass presentation, or even the reduction of the main function of national governments to managing the state in the interest of international capital. Broadly speaking, new technologies transform the relationship between the public and the private. However, what is particularly relevant in *Specters of Marx* is Derrida's argument that the speed and pervasiveness of new media oblige us more than ever to think about the virtualization of space and time, and prevent us from opposing an event to its representation, and 'real time' to 'deferred time'.[9] Moreover, not only do tele-technologies deconstruct the political space, but they also destabilize and deconstruct the perception of the human being as separate from his tools and as a master of them. Clark writes:

To conceive technology as a prosthesis that alters the very nature of its seeming user is to be reminded how technical inventions have always been in excess of their concepts, productive of unforeseeable transformations. This deconstruction of the Aristotelian system enables each invention to be seen as an irruption of the other, the unforeseen disrupting the very criteria in which it would have been captured.[10]

Clark's comment clarifies that there are certain effects of technological innovation that are beyond ('in excess of') what can be expected within (and produced by means of) a procedural method. This leap beyond the programmable, which was once ascribed to 'genius' or 'inspiration' and therefore recuperated as part of the philosophy of subjectivity, is now the result of deconstruction as the bringing about of what cannot be calculated or programmed.[11] As a striking example of such unexpected consequences of technology we could perhaps think of the Internet and its growth, out of a seminal military structure of the 1970s (the ARPANET), into an extended network whose expansion is only in part controllable. Derrida names such a leap beyond the programmable 'the coming of the other.'[12] In order to understand this concept better, one must recall that—as Clark explains—'[d]econstruction represents a rejection of the view, powerful since the seventeenth century, that humanity necessarily understands what it has made better than what it has not.'[13] This observation posits a fundamental relationship between deconstruction and technology. Both Clark and Derrida acknowledge that technology has the capacity for interrupting reason and calculation, and for bringing forth something unexpected that cannot be totally understood within the existing conceptual framework. For instance, tele-technologies challenge the very concept of 'invention' as something that 'is assignable to a human subjectivity, individual or collective, responsible for the discovery or production of something new and publicly available'.[14] The concept of invention must therefore be reinvented, and it exceeds the recognizability of the producer and of the product.[15]

In sum, according to Derrida, today's technology is 'in deconstruction' because, with its capacity to generate unforeseen consequences, it challenges received notions of invention and instrumentality, and in so doing it challenges the (Aristotelian) notion of 'tool' within which technology itself has been constricted for centuries. This is also the reason why, as Clark makes explicit, technology cannot be simply an 'object' of deconstruction understood as a discrete methodology of analysis or critique: '[t]echnology is "in de-

construction" as the condition of its being: for an originary prosthetic supplemementarity or originary technicity both opens and prevents realization of the Aristotelian and techno-scientific concepts
of technology.'[16] Accordingly, as I have noticed earlier on, Derrida
argues that deconstruction is something that 'happens' within a
conceptual system, rather than a methodology, and that any conceptual system is always in deconstruction, because it unavoidably
reaches a point where it disassembles its own presuppositions. And
yet, since it is possible to remain oblivious to the permanent occurrence of deconstruction, there is a need for us to actively 'perform'
it. Therefore, although what Derrida terms 'practical deconstruction' is already occurring in new technologies, there remains a need
to make such occurrence visible through an active investigation of
new technologies' 'complicity' with instrumentality. But to what
extent and in what way does software both 'open and prevent the
realization of instrumentality'—as Clark would have it—that is, of
the traditional philosophical understanding of technology as a tool?
To what extent and in what way does software exceed its own instrumentality by giving rise to unforeseen consequences? And most
importantly, if technology and writing have been subjected to the
same process of devaluation in the history of Western thought because of their alleged instrumentality, to what extent a deconstruction of software involves a deconstruction of its relationship with
writing and language, and in what way can this deconstruction
actually be performed?

In order to expand on this point, let me now turn to the analysis
of the relationship that software entertains with Western thought on
the one hand and with writing on the other offered by Katherine
Hayles in her 2005 book, *My Mother Was a Computer*. Hayles's argument is worth examining at length because her understanding of
software has had a lasting influence on the field of Software and
Code Studies.[17] Hayles reframes the problem of the relationship
between software and instrumentality in terms of the relation between 'code' and 'metaphysics', at the same time introducing the
concepts of language and writing as significant terms of this relation. Hayles's understanding of 'code' is not identical to my understanding of 'software', although they can be considered equivalent
up to a certain point. For Hayles, code is 'the language in which
computation is carried out'—whereby 'computation' is defined as 'a
process that starts with a parsimonious set of elements and a relatively small set of logical operations', which, instantiated into some
kind of material substrate (such as a computer), can build up in-

creasing levels of complexity, 'eventually arriving at complexity so deep, multilayered, and extensive as to simulate the most complex phenomena on earth, from turbulent flow and multiagent social systems to reasoning processes one might legitimately call thinking.'[18]

Hayles defines metaphysics as a model 'for understanding truth statements' which is 'so woven into the structure of Western philosophy, science, and social structures that it continually imposes itself on language and, indeed, on thought itself, creeping back in at the very moment it seems to have been exorcised.'[19] She is particularly interested in examining the relationship between code and 'the metaphysics of presence' which, after Derrida, she defines as the metaphysical yearning for the 'transcendental signified'—that is, for 'the manifestation of Being so potent that it needs no signifier to verify its authenticity', or, in Derrida's terms, 'a *concept signified in and of itself*, a concept simply present for thought.'[20] According to Hayles, code has a very loose relationship with metaphysics (or with ontology, which is another name she gives to the metaphysics of presence and to the transcendental signified). She goes as far as to say that code 'inherit[s] little or no baggage from classical metaphysics', because it reduces ontological presuppositions to a minimum.[21]

The question of the relations between metaphysics and code is considered by Hayles in the broader context of the relations between speech, writing, and code. She establishes a distinction between natural language on the one hand and programming languages and code on the other. She identifies a number of differences between the two: for instance, code is addressed both to humans and to computers, it is developed by small groups of technicians, and it is integrated into commercial production cycles and capitalist economics. She notices: '[a]lthough virtually everyone would agree that language has been decisive in determining the specificity of human cultures and, indeed, of the human species, the range, extent, and importance of code remain controversial.'[22] On the one extreme of the spectrum there are people who see code as a niche artificial language aimed at computers. On the other extreme, we can find scientists such as Stephen Wolfram and Harold Morowitz, who advance the hypothesis that the whole universe is fundamentally computational, therefore viewing code as the language of nature—that is, as the *lingua franca* of all physical reality. Whether we view code as something very specific or utterly pervasive, Hayles urges for a theoretically sophisticated understanding of the interactions between code and language. She points out that, while there exists an enormous body of literature dealing with human languages and programming languages separately, scholarship engaging with the relationship between the two is comparatively limited. Yet interactions between language and code pervade our world. For example, human communication over the Internet involves natural language (for instance when writing an email or using a chat line) as well as code (for instance protocols of communications between networked computers). Interactions between natural language and code take place nearly every time computers are relied upon in order to perform everyday tasks. 'Language alone', Hayles writes, 'is no longer the distinctive characteristic of technologically developed societies; rather, it is language plus code.'[23]

Up to this point it looks like Hayles's main preoccupation is the relationship between natural language and code. In fact, in the rest of the book she rapidly slips into a triadic model which involves speech, writing, and code. Hayles identifies throughout history three main 'discourse systems'—namely, the system of speech, the system of writing, and the system of digital computer code—all of which are still at work in contemporary culture.[24] Each of these systems is associated with a specific 'worldview'—that is, a particular set of premises and implications which can be detected, exemplarily and respectively, in the semiotic theory of Ferdinand de Saussure, in the grammatological thought of Jacques Derrida, and in the theories of a number of thinkers such as Wolfram and Morowitz, but also Ellen Ullman, Matthew Fuller, Matthew Kirschenbaum, and Bruce Eckel, who have dealt with programming languages.[25] According to Hayles, the worldviews of speech, language, and code are connected to each other in a complex process of exchange and interplay which she calls 'intermediation'.[26] The investigation of the processes of intermediation between speech, writing, and code assumes an understanding of the differences and similarities between the three worldviews associated with them. In turn, such differences become mostly apparent precisely in the relationship that these three worldviews entertain with metaphysics.

Thus, we are brought back to the initial question of the relation between code and metaphysics. We have already seen that for Hayles computation must be understood in a much broader sense than as something that goes on within a computer. In this view, computation is not limited to digital machines and binary code, but can form the basis for the physical universe. She refers to Stephen Wolfram, who, in his book *A New Kind of Science*, makes the claim that the whole universe can be explained through processes of computation and according to the theory of cellular automata. For Wolfram, the complexity of the universe can be viewed as emerging through successive cycles of computation, starting from binary elements. Complexity emerges not from the variety of the starting elements but from the number of iterations and from the unpredictability of their results.[27] Hayles does not embrace the theory of the computational universe as a plausible explanation of physical reality, but treats it as a productive hypothesis that provides her with a means to articulate the worldview associated with code.[28] As we have seen, for her, the worldview of computation reduces ontological requirements to a minimum. She explains:

> Rather than an initial premise (such as God, an originary Logos, or the axioms of Euclidean geometry) out of which multiple entailments spin, computation requires only an elementary distinction between something and nothing (one and zero) and a small set of logical operations. The emphasis falls not on working out the logical entailments of the initial premises but on unpredictable surprises created by the simulation as it is computed.[29]

Thus, for Hayles the distinction between zero and one minimizes the need for metaphysical foundations and the computational view of the universe requires no ontological foundations, other than the minimal presuppositions needed to set the system running. 'Far from presuming the "transcendental signified" that Derrida identifies as intrinsic to classical metaphysics', she remarks that 'computation privileges the emergence of complexity from simple elements and rules.'[30]

This last part of Hayles's argument seems to me rather problematic. In fact, when discussing the theory of the computational universe, Hayles actually detects its fundamental ambiguity—that is, its uncertainty whether computation should be understood as a metaphor (however pervasive it is in our culture) or as having an 'ontological status as the mechanism generating the complexities of physical reality.'[31] She explicitly avoids taking sides on such a controversial issue (especially since the theory of the computational universe has not been proved). Rather, she accepts that such a question remains undecidable as of today and that the computational universe is able to function simultaneously as means and metaphor. She argues that, regarded as a culturally potent metaphor, computation can actually invest the social construction of reality, as is the case of the reorganization of the US military according to 'network-centric warfare'. Here the presupposition of information as a key military asset leads to the reorganization of the military as a mobile and flexible force and as 'a continuously adapting ecosystem.'[32] Thus, Hayles continues, '[a]nticipating a future in which code (a synecdoche for information) has become so fundamental that it may be regarded as ontological, these transformations take the computational vision and feed it back into the present to reorganize resources, institutional structures, and military capabilities. Even if code is not originally ontological, it becomes so through these recursive feedback loops.'[33] In this—again debatable—statement, Hayles seems to have moved away from the conception of information that she supported in *How We Became Posthuman*. In that book she showed how Shannon and Weaver's understanding of information as a mathematical function independent of its material substrate was able to generate the possibility of ultimately thinking the human as a disembodied entity. In *My Mother Was a Computer*, conversely, she analyzes the formal definition of computation in order to explore its potential to generate embodied social and cultural (as well as physical) structures. In *How We Became Posthuman*, she regarded Hans Moravec's desire to upload the human consciousness into a computer as emblematic of a disembodied posthumanity, which she argued against. In *My Mother Was a Computer*, she embraces Richard Doyle's observation in *Wetware* that the desire to upload one's consciousness into a computer (and thereby achieve immortality) has already provoked a new perception of one's self and of the others, or what Doyle calls a 'techno-social mutation.'[34]

What I want to emphasize here is that, ultimately, Hayles's theory of computation does not seem to break free from the dilemma of whether code has 'ontological status'. Even more importantly, I want to argue that she cannot get rid of it precisely because she conceptualizes code within a triadic model that sets speech, writing, and code as three separate entities associated (as we have seen above) with three separate worldviews that, albeit in complex interactions (which she calls intermediations), can still be placed in a kind of temporal progression with respect to their relationship with metaphysics. I believe that this triadic model needs to be rethought in order to give account of the relationship between 'code' and 'metaphysics', as well as of the one between technology (and specifically software) and instrumentality. Let me now clarify this point by returning to the fact that Hayles associates a specific reference text—and a specific theory—with each of the three worldviews,

which is presumed to give an account of that worldview's relationship with metaphysics.

According to Hayles, the worldview of computation shifts the
locus of complexity from Logos (that is, from the ideal originary
point that both exceeds and generates the world) to 'the labor of
computation that again and again calculates differences to create
complexity as an emergent property of computation.'[35] Here Hayles
acknowledges the importance of Derrida's deconstruction of metaphysics, and yet she claims that Derrida's grammatology locates
complexity in the 'trace', and that therefore it cannot be relied upon
in order to give an accurate account of the processes of computation. Actually, for her, the different locations of complexity in Saussurean linguistics, Derridean grammatology, and computation
'have extensive implications for their respective worldviews'.[36] She
starts from the examination of the foundational text of modern linguistics, Ferdinand de Saussure's *Course in General Linguistics*. According to Saussure, the sign has no 'natural' or inevitable relation
to that which it refers to—that is, in his own words, 'the linguistic
sign is arbitrary'. He regards speech as the truth of the language
system (*la langue*) and writing as merely derivative of speech and
explains: '[a] language and its written form constitute two separate
systems of signs. The sole reason for the existence of the latter is to
present the former. The object of study of linguistics is not a combination of the written word and the spoken word. The spoken word
alone constitutes that object.'[37]

Hayles follows Derrida's critique of Saussure's prioritization of
speech over writing. And yet, her main critique to Saussure is that
he tends to underplay the role of 'material constraints' of any kind
in the functioning of the sign. Briefly put, Hayles's argument is that
the arbitrariness of the linguistic sign is much less valid in code than
in speech and writing. She emphasizes that material constraints
play a productive role in computation, 'functioning to eliminate
possible choices until only a few remain.'[38] Saussure ignores such
constraints, and views meaning as emerging only from differential
relations between signs. As Jonathan Culler remarks, for Saussure
signs are 'purely relational or differential entities'.[39] On the
contrary, for Hayles, although material constraints also work in
speech and writing, they are much more important in code. While
Saussure's theory erases materiality, the development of computers
has been characterized by a dramatic centrality of it, from the need
to cool off vacuum tubes in the computers of the 1950s to concerns
over the limits of the miniaturization of silicon-based chips. 'For

code', Hayles concludes, 'the assumption that the sign is arbitrary must be qualified by material constraints that limit the ranges within which signs can operate meaningfully and acquire significance. . . . In the worldview of code, materiality matters.'[40]

In sum, for Hayles, code has a much stronger connection to materiality than speech and writing. When discussing the 'worldview of writing', she applies an analogous kind of criticism to Derrida — namely, that grammatology does not take into account the role of materiality in the process of signification. Here Hayles's argument becomes problematic again and deserves a careful examination. In fact, not only do I want to argue that Derrida's conception of writing is based on his recognition of the materiality of the sign. Furthermore, as I began to show in Chapter 1, grammatology has much wider implications than Hayles recognizes, since it constitutes a problematization of the whole of Western metaphysical thought. Actually, I want to suggest that Derrida's understanding of writing can lead to the questioning of Hayles's own triadic model as well as to a radical reframing of her 'ontological dilemma' and, even more importantly, to an understanding of 'writing' that is particularly helpful for the conceptualization of what I name 'software' (and Hayles names 'code').

Her discussion of Saussure's semiotics leads Hayles to ask whether any processes of signification actually take place in computation, and, if so, what a sign in the Saussurean sense (that is, as the unity of a signifier and a signified) would become in code. She follows Kittler's observation from his essay 'There Is No Software' that ultimately everything in digital computers is reduced to changes in voltages, and advances the proposition that signifiers in code coincide with voltages, while signifieds are 'interpretations that other layers of code give these voltages.'[41] In other words, Hayles views the microcircuitry in a digital computer as a signifier, and the first layer of code (namely, sequences of binary digits) as its signified. In turn, a higher level of code treats such sequences of binary digits as a signifier and attributes a certain meaning to them, according to the rules of the programming language in which that code is written. Programming languages operating at still higher levels translate the lower levels of signification into commands that more closely resemble natural language. Hayles explains that '[t]he translation from binary code into high-level languages, and from high-level languages back into binary code, must happen every time commands are compiled or interpreted, for voltages and the bit stream formed from them are all the machine can understand.'[42]

What is worth noting in this passage is that Hayles views the correspondence between circuitry and binary digits as a process of translation in which materiality plays a very important part. For her, in
the computer the chain of zeroes and ones eventually 'becomes'
circuitry. She argues further:

> Hence the different levels of code consist of interlocking chains of
> signifiers and signifieds, with signifieds on one level becoming
> signifiers on another. Because all these operations depend on the
> ability of the machine to recognize the difference between one
> and zero, Saussure's premise that differences between signs
> make signification possible fits well within computer architec-
> ture.[43]

That computation partly fits a Saussurean model should come as
no surprise, since historically programming languages have been
developed according to structuralist linguistics. As I will show in
greater detail in Chapter 5, the modern theory of formal languages
(which is concerned with the specification and manipulation of languages, be they natural languages such as English or programming
languages such as Pascal) was originated mainly by the work of
Noam Chomsky, who in the 1950s, at MIT, developed a model for
the formal description of natural languages based on what he called
'replacement rules' and 'transformations'.[44]

However, the relevant point here is that Hayles invokes a Saussurean understanding of the sign throughout her analysis of code,
while at the same time upholding a stronger account of materiality
within the Saussurean framework. In order to give an account of the
materiality of code, Hayles follows Alexander R. Galloway in
understanding code as executable. For Galloway, this is the essential difference between code and any other language. This is also
why code is 'so different from mere writing': '[c]ode is a language,
but a very special kind of language. Code is the only language that
is executable.'[45] Hayles relies on Galloway's statement, as well as on
John Austin's theory of speech acts, in order to substantiate her
assertion that code has a 'performative' nature, albeit in a stronger
sense than language. As it is widely known, Austin's theory establishes the existence of declarative sentences that do not describe
anything, and of which it makes no sense to ask whether they are
true or false.[46] Austin points out that the utterance of these sentences coincides instead with the 'doing' of an action that cannot be
described as the action of saying something. The classical example
given by Austin is the 'I do' uttered in a marriage ceremony. Such

utterances he names 'performatives' (as distinct from 'constatives', which are descriptive). The difference lies in the fact that, while constatives can be true or false, performatives can only be 'happy' or 'unhappy'. A performative is happy depending on the conditions under which it is uttered: normally, there must be some accepted conventions regarding the performative that are understood by the participants in the conventional procedure, which in turn has to be carried out correctly and generate the expected social behaviour in the participants themselves.[47] Quite clearly, the 'felicity conditions' of performatives unfold in the social realm. Accordingly, Hayles stresses that, while performative language causes changes 'only' in the mind and behaviour of people, code always has a very particular involvement with materiality, since 'it causes things to happen, which requires that it be executed as commands the machine can run.'[48] In other words, for her, the materiality of code is more straightforward than the materiality of performative utterances. 'Code that runs on a machine is performative in a much stronger sense than that attributed to language.'[49] In fact, while the performative force of language is 'tied to external changes through complex chains of mediation', by contrast,

> code running in a digital computer causes changes in machine behaviour and, through networked ports and other interfaces, may initiate other changes, all implemented through transmission and execution of code. Although code originates with human writers and readers, once entered into the machine it has as its primary reader the machine itself. Before any screen display accessible to humans can be generated, the machine must first read the code and use its instruction to write messages humans can read. Regardless of what humans think of a piece of code, the machine is the final arbiter of whether the code is intelligible. If the machine cannot read the code or if the program does not work properly, then the code must be changed and corrected before the machine can make things happen.[50]

This passage presents a number of interesting points. First of all, Hayles introduces the terms 'writing' and 'reading' in relation to code, without substantially questioning them. But what would it actually mean for a computer to 'read' code? And what does it mean, exactly, for a human being to 'write' it? Moreover, while marking out the difference between performative language and performative code, the above passage also introduces a problem of agency—that is, it clarifies that, as far as code is concerned, computers are the arbiters of the competence of the utterance. Hayles already articulated this very point in her 1999 book, *How We Became Posthuman*, when she observed that while '[i]n natural languages, performative utterances operate in a symbolic realm, where they can make things happen because they refer to actions that are themselves symbolic constructions, actions such as getting married, opening meetings, or as Butler has argued, acquiring gender', programming languages operate 'the act of attaching significance to . . . physical changes.'[51]

However careful these observations, Hayles's understanding of materiality and of the relationship between materiality and code remains questionable. Importantly, she describes the materiality of code as more 'direct', more straightforward than the materiality of language and writing. On the contrary, in her rereading of Butler, the performativity of gender is exemplary of a 'mediated' performativity—a performativity that has effects mainly in the 'symbolic' realm. An analogous assimilation of Derrida and Butler's concepts of performativity as limited to the 'symbolic' or 'discursive' realm comes back from time to time in the theories of technology and gender.[52] And yet I believe that these critiques underplay the role of materiality in Butler's theory of performativity. As Butler herself remarks in *Excitable Speech*, language can produce very powerful effects on the materiality of the human body—effects that Hayles would without doubt consider quite 'straightforward'. To give but an example, in her poignant analysis of hate speech, Butler argues that human beings can be injured by language, and that linguistic injuries such as racial invectives can produce physical symptoms such as blushing and rage.[53] On the other hand, on what grounds can Hayles claim that code has a 'direct' relation with materiality? When stating that code causes changes in machine behaviour 'through networked ports and other interfaces', she actually acknowledges the mediated character of materiality in code.

What I want to suggest here is that the famous rereading that Derrida gives of Austin's theory in his 1972 essay, 'Signature, Event, Context', can be extremely helpful in order to think about the materiality of code in an alternative way.[54] Famously, in this work, Derrida calls into question the traditional understanding of 'communication' as a vehicle that circulates an identifiable content—that is, a meaning. The meaning or content can be 'communicated' by different technical means—namely, 'by more powerful technical mediations'—for example, by writing—without being affected.[55] Again, this interpretation of writing as something that simply 'extends' the domain of communication is proper to philosophy. Derrida exemplifies this understanding of writing through Condillac's *Essay on the Origin of Human Knowledge*, where writing is defined as a means of making one's thought known to persons who are 'absent'.[56] According to Derrida, this notion of absence is not specific to writing; in fact, it is characteristic of every sign. He explains:

> In order for my 'written communication' to retain its function as writing, i.e., its readability, it must remain readable despite the absolute disappearance of any receiver, determined in general. My communication must be repeatable—iterable—in the absolute absence of the receiver or of any empirically determinable collectivity of receivers.[57]

In other words, a piece of writing must be repeatable in order to function *as writing*. The capacity of writing to remain writing in the absence of its intended reader and/or producer—this iterability—constitutes the capacity of the written sign to break free from its context (that is, the 'empirically determinable collectivity' of receivers or producers). For Derrida, this capacity of functioning after having been severed from their context pertains to every kind of sign. He states that '[e]very sign, linguistic or non-linguistic, spoken or written (in the current sense of this opposition), in a small or large unit, can be *cited*, put between quotation marks; in so doing it can break with every given context, engendering an infinity of new contexts in a manner which is absolutely illimitable.'[58] At this point it becomes clear why, for Derrida, the structure of writing is the structure of every possible mark. The possibility of disengagement from a context and of citationality belongs to the structure of every mark, spoken or written—otherwise it could not function as a mark. 'What would a mark be that could not be cited?' Derrida asks.[59]

It is from this perspective that Derrida approaches the problematic of the performative. Austin seems to consider the performative as something that does not describe something preexistent (something outside of language and prior to it); rather, the performative 'produces or transforms a situation, it effects'. For this reason, Austin was obliged to free the analysis of the performative from what Derrida calls the 'authority of the truth value' and to substitute for it 'the value of force'.[60] So, has Austin shattered the traditional concept of communication? For Derrida, he has not. In fact, Austin's performative always requires a value of context—that is, the intentionality of the speaker, the communication of an intentional meaning. In Derrida's words, Austin has not 'interrogated infelicity as a law'. Derrida thus asks '[w]hat is a success when the possibility of infelicity . . . continues to constitute its structure?'[61] In order to prove this point, and to perform at the same time such an 'interrogation of infelicity as a law', Derrida criticizes Austin's understanding of citationality. While Austin excludes the possibility of a performative being quoted and actually considers a quoted performative (such as the pronunciation of the nuptial 'I do' by an actor on stage) as 'abnormal, parasitic', Derrida objects that all language is citational in order to function as such, and *especially* a performative. A performative is only possible if it is citational: how could the 'I do' function if it were not identifiable as the citation of the marriage formula? For Derrida '[w]hat Austin excludes as anomaly . . . is the determined modification of a general citationality—or rather, general iterability—without which there would not even be a "successful" performative.'[62]

If analyzed from this point of view, Hayles's distinction between the performativity of language and the performativity of code would not hold. In fact, for Derrida, every process of signification requires the sense of the material persistence of the sign, and there is no opposition between performatives operating in 'the symbolic realm' and performatives that take place in computer processors. Actually, Hayles's analysis of code brings back a certain separation between materiality and signification that I argue is not particularly helpful in understanding how code works. For instance, drawing on such a distinction, Hayles isolates two more characteristics of code: besides being material and executable, code is also hierarchical and discrete. The hierarchical structure of code seems to depend precisely on its proximity to materiality. The lower the level, the closer code comes to the simplicity of zeros and ones. The higher the level, the more similar to natural language code becomes. Hierarchy is closely related to discreteness, which in turn can be understood as the very fact that code is digital. Simply put, digitization is the operation 'of making something discrete rather than continuous, that is, digital rather than analog'.[63] For Hayles, crucially, digitization is a specific characteristic of code that can hardly be found in speech and writing. She writes:

> The act of making discrete extends through multiple levels of scale, from the physical process of forming bit patterns up through a complex hierarchy in which programs are written to compile other programs. Understanding the practices through which this hierarchy is constructed, as well as the empowerments and limitations the hierarchy entails, is an important step in theorizing code in relation to speech and writing.[64]

Hayles's general argument is that in the progression from speech to writing to code, every successor regime introduces new features that we cannot find in the predecessor, and that this is the case for discreteness. She acknowledges that for Derrida, spacing was what made writing not a simple transcription of speech but something that exceeded speech. In the same way—she states—code exceeds both speech and writing, and cannot be incapsulated in them. Another feature that cannot be found in speech or in writing is compiling—that is, the process of translation of high-level code into binary digits. Compilers are necessary if code has to become operative. Hayles relates such specificity both to the process of digitization and to the fact that code implies a partnership between humans and machines, in which the act of compiling is carried out both by machines and humans. As a conclusion, Hayles suggests that the process of digitization has 'ideological implications'.[65] She draws here on Wendy Hui Kyong Chun's assertion that 'software is ideology' and on her Althusserian reading of computer interfaces.[66] Drawing on Althusser's understanding of ideology as the subject's imaginary relationship to his or her real conditions of existence, Chun claims that computer interfaces (including basic desktop metaphors such as folders and trash cans) create an imaginary relationship of the user to 'the actual command of the core machine', that is, to the 'real' technical context within which the user's actions are actually given meaning and responded to. Following Chun, Hayles speaks of a disciplining of the user by the machine to become a certain kind of subject.[67]

Although Hayles's argument is utterly accurate, I believe that it does not do justice to the practice of 'discreteness' taking place in code. I actually want to argue that what Hayles calls 'discreteness' can again be seen as a characteristic of *every* sign. For instance, the emergence of alphabetic writing can be seen as a process of 'making discrete' in Hayles's terms. As I showed in Chapter 1 following Derrida's rereading of Leroi-Gourhan's work, alphabetic writing is the result of a process of 'linearization' that transforms 'picto-ideography'—that is, the early form of graphism tightly associated with figurative art and independent from spoken language—into a sequence of phonetic symbols subordinated to spoken language and to its linear temporality. Before examining this important point further, it is worth recapitulating Hayles's argument so far.

In sum, Hayles basically depicts the relationship between 'code' and 'metaphysics' as a loose one, since code is characterized by the minimization of its ontological premises. And yet, she seems to continue interpreting code through the structuralistic version of linguistics (Saussure, Austin) that, albeit partially explaining the functioning of code, does nothing to further our understanding of how software exceeds instrumentality. Moreover, she establishes a relationship of 'intermediation' between code and writing, adding 'speech' as the third term of the triadic model that is supposed to account for the development of contemporary technology. In fact, she differentiates between language and writing, on the one hand, and code on the other by arguing that the latter has a stronger relation with materiality. In doing so, she seems to bring back the same distinction between the material and the symbolic that we have seen as the foundation of Western thought and of the philosophical (in Hayles's words, 'metaphysical') devaluation of technology (and writing).

Importantly, the tradition of originary technicity calls into question precisely such distinction between the material and the symbolic, thus also questioning the instrumental understanding of technology. What I want to suggest here is that, rather than rethinking the relationship between 'code' and 'metaphysics' (in Hayles's words) as a minimization of the ontological requirements of code—a conceptual move that ultimately seems to trap Hayles's argument in an ontological dilemma—it would be more productive to approach the problem from the point of view of originary technicity. As I have shown at the beginning of this chapter, such an approach opens up the following questions: first, In what way does software both participate in instrumentality at the same time as exceeding it? and secondly, To what extent can the relationship between technology and writing (as both traditionally excluded by philosophy as 'instrumental' *and* constitutive of it) help us answer the first question?

With regard to the relation between software and instrumentality, Hayles's argument goes to great lengths to clarify how software works, but—as I have remarked above—falls short of taking into consideration software's potentiality for producing unexpected consequences that go beyond its 'normal' functioning. And yet, I want to suggest that the circumstances in which software does *not* function—better, in which it does not function *as expected*—could tell us more about software than those in which software 'works'. But in order to understand such circumstances, a different approach is needed—namely, an approach that views malfunctions as revealing points (or points where the conceptual system underlying software is clarified) rather than just seeking to explain how software functions ordinarily.

If examined in relation to the second question—that is, what the relationship between writing and technology means in the framework of originary technicity—Hayles's argument clearly acknowledges the need for an understanding of the relationship between 'code' (which I still regard here as a synonym for 'software') and writing. However, her attempt to place code and writing in a unitary framework does not give an account of the role that materiality plays in writing itself, and ultimately in software. Let me start from this second point and draw on Derrida's grammatological thought in order to investigate the role of materiality in software further, thus clarifying my own understanding of the relationship between 'code' (or software) and writing. In turn, this analysis will prove useful for showing how software participates in and exceeds instrumentality. Moreover, it will also confirm that Hayles's call for a 'turn to materiality', which has had great relevance in media studies, is in fact quite belated, and that, as I have already hinted above, such a call actually results from a misreading of the poststructuralist tradition Hayles draws on (but ultimately departs from).[68]

As we have seen earlier on, for Derrida the subordination of writing to speech has meaning only within the system of Western metaphysics, whose premises have been inherited by human sciences and particularly by linguistics. In Richard Beardsworth's terms, for Derrida, 'the theory of the sign is essentially metaphysical.'[69] As is widely known, in *Of Grammatology* Derrida focuses on the deconstruction of linguistics from this viewpoint. He takes the deconstruction of linguistics and of its central concept, the sign, as exemplary for the deconstruction of metaphysics, not least because the sign is exemplary of the metaphysical devaluation of materiality. The signifier is a material entity, such as a sound or a graphic sign; the signified belongs to the realm of concepts. For Derrida, this opposition is the foundation of all the other oppositions that characterize Western metaphysics (infinite/finite, soul/body, nature/law, universal/particular, etc.). Therefore, deconstructing the sign is a fundamental move precisely because, as Beardsworth explains in *Derrida and the Political*, 'metaphysics is derived from the domination of a particular relation between the ideal and the material which assumes definition in the concept of the "sign."'[70] The sign constitutes the foundation of the distinction between signifier and thing, a distinction which in turn is the basis of *episteme* and therefore of metaphysics. Again in Beardsworth's words, 'metaphysics constitutes its oppositions (here: the non-worldly/worldly and the ideal/material) by expelling into one term of the opposition the very possibility of the condition of such oppositions.'[71] To clarify this important point further, let me now examine Derrida's analysis of the way in which such expulsion is performed in Ferdinand de Saussure's thought at some length.

Saussure argues that linguistics (as a science of language) must exclude from its objects of study the graphic sign. In fact, linguistics as a discipline is based for Saussure on the very separation of the abstract system of *langue* from the empirical multiplicity of languages with all their variations. This abstraction depends in turn on 'a distinction between what is internal and essential, and what is external and accidental to the system of *langue*'.[72] In *Of Grammatology*, Derrida detects some fundamental inconsistencies in Saussure's view. First of all, in spite of being conventional, or 'a pure institution', *langue* for Saussure is based on the 'natural unity' between its two components—that is, meanings and what he calls sound-images (*l'image acoustique*).[73] In other words, Saussure claims that there is a 'natural' correspondence between the signifier and its meaning, or signified. The phonetic pronunciation of a word is somehow more 'natural' than its written form. As Beardsworth points out, 'it is this natural order that allows Saussure to set linguistics up as a science'—that is, Saussure seeks a foundation for linguistics in a supposed natural order of things.[74] On the contrary, Derrida argues that the exclusion of writing from linguistics must be viewed as an 'ethico-theoretical decision'.[75] Such a decision is masked under the apparent naturality of the object under consideration, but 'revealed by the obsessive insistence with which the founder of linguistics wishes to expel writing from the essence of language'.[76] What is significant is not merely the devaluation of writing. Rather, such devaluation shows that linguistics has been founded on a double movement: first, the making of a decision, and secondly, the justification of such decision through the claim that it is natural—therefore, a disguise of the decision itself under the pretence of naturality. For Derrida, both linguistics and philosophy are predicated on the normative exclusion of writing from truth. However, this exclusion is the cause of a number of contradictions in Saussure's thought.

The first contradiction is located by Derrida in the natural hierarchy that Saussure poses between speech and writing. For Saussure, the sign is arbitrary—therefore, it is non-natural, a social institution. But to say that there is a natural subordination of writing to speech whilst maintaining that all signs are arbitrary is ultimately contradictory. Furthermore, Saussure explicitly states that the word is a pragmatic decision, clearly admitting that to take the word as the minimum unit of analysis is a decision made at the foundation of linguistics.[77] Thus, he practically removes the objectivity of linguistics while instituting it as science. What Derrida terms 'an ethico-theoretical decision' is precisely the movement that institutes the object of a 'science' but pretends to be natural, whereas (being a decision) it is not—and for this reason Derrida argues that such a decision is 'violent'. Beardsworth comments: 'The irreducibility of a decision shows that the most innocent "theorist" is always also a legislator and a policeman. It is in this sense that any statement is a judgement which carries "political" force.'[78] This passage clarifies very well why every theory is ultimately political. There is always violence in theory, and every theorist is a legislator insofar as every theory or discipline needs to police its boundaries. Derrida is not suggesting that we avoid decisions, or that all decisions are equally violent. The point is rather that a decision is always needed, but not all decisions are the same: some of them recognize their legislative force, while others disguise it under a claim to naturality (affirming that they are 'objective science'). Moreover, 'the acknowledgement of the prescriptive force of one's statements' may make one more ready to transform a disciplinary field, given that the field is not a natural representation of a preexisting 'real'. This is what Derrida terms the argument of a 'lesser violence' in 'a general economy of violence.'[79] 'Lesser violence', at the level of theory, means acknowledging the normative force of one's decisions—that is, the fact that such decisions shape the field, the theory, the discipline.[80]

However, according to Derrida, the force of Saussure's decision is revealed in what it suppresses, because what is suppressed returns as a contradiction, and causes either repeated acts of violence—that is, the policing of the field—or a radical reinscription of Saussure's legislative decision. According to Beardsworth's exemplary exegesis of Derrida's thought, Derrida accomplishes such reinscription of structural linguistics in two steps. The first step is his famous generalization of writing, which actually finds its basis in Saussure's own theory. In fact, Saussure argues that writing covers the whole field of linguistics, since the sign in itself is 'immotivated', and writing is exemplarily immotivated. In other words, writing, being entirely conventional, perfectly exemplifies the conventionality of language—and all speech is already writing in its being immotivated. For Derrida, writing (the *graphie*) 'implies the framework of the *instituted trace*, as the possibility common to all systems of signification'.[81] The concept of 'instituted trace' is very important here, since it represents the possibility of making conceptual distinctions. Thus, the instituted trace represents the moment that precedes the opposition between nature and convention, allowing for the very possibility of their separation. The concept of the instituted trace takes into account and comprehends Saussure's act of foundation of linguistics as a discipline, with its refusal to include writing. As we have just seen, there is a strong relationship between the instituted trace and disciplinarity, since the former accounts for the foundation of a disciplinary space with its constitutive exclusions, as well as for the return of that which is excluded within the disciplinary space.

A further step on the investigation of what Derrida calls the 'economy of violence'—a step that corresponds to another contradiction in Saussure—concerns the *phone,* and is the most important reinscription of Saussure made by Derrida. Here Derrida confronts the problem of materiality straightforwardly. He reads Saussure against himself once again, this time in order to show that 'both philosophy and linguistics are derivatives of a movement which constitutes them, but which they disavow in order to appear as such.'[82] His argument turns around the difference that Saussure recognizes between phonemes and their 'sonorous concretization'—that is, between the 'sound-image' and its materialization. Such difference allows, for instance, for multiple pronunciations of the same phoneme, which are at the same time recognized as pronunciations of the same phoneme. The pivotal passage is Saussure's use of writing as a metaphor for the reduction of the phonic substance.

Saussure uses writing as an example for phonetics, and states that the sign is arbitrary, negative, and differential—that is, it functions only through (reciprocal) opposition, and the means by which it is produced are unimportant.[83] There is a difference between the materialization of each phoneme and the acoustic-sound 'which they presuppose in order to be recognized as such, whatever the form of their materialization'.[84] This difference is transformed, in Platonism, into the difference between the transcendental and the empirical, the ideal and the material. This very difference constitutes consciousness and founds the possibility of recognizing things. One must be reminded at this point of Derrida's understanding of consciousness, which in turn is informed by his rereading of Husserl's phenomenological thought.[85] Husserlian phenomenology separates ideal objects, attained through what Husserl calls 'phenomenological reduction', and the world, with its empirical variations. In *The Origin of Geometry*, Husserl derives the possibility of phenomenological reduction from writing. For him, writing constitutes ideal objects because the condition of their ideality is precisely 'their repetition through time and space', which in turns 'depends on their inscription on a support which transcends the empirical context'.[86] According to Derrida's radicalization of Husserlian phenomenology, the transcendental is always impure, always already constituted through materiality (the empirical), because the condition of consciousness is repetition. The consequences of Derrida's understanding of the relationship between the transcendental and the empirical are of great relevance for the conceptualization of writing, because they lead to the conclusion that writing and the material are not opposed. In fact, writing *is* material because materiality is the condition for writing itself, and for signification.

The difference examined by Derrida is between an empirical *t* and our ability to recognize it as an instance of the letter *t*. Derrida notes that the sound-image 'is not a phonic sound object but the "difference" of each of its concretisations, that is to say it is the possibility or schema of each of its materializations'.[87] In other words, to make the phoneme recognizable we need to be able to relate it to other phonemes, but this relation is only possible through its inscription in the empirical; therefore, the transcendence of an empirical sound is possible only via the empirical repetition of it, which needs to be transcended to make it recognizable. The phenomenological reduction of materiality makes the materialization of the 'sound-image' possible. Derrida writes:

> The sound-image is the structure of the appearance of the sound which is anything but the sound appearing . . . not the sound being heard but the being-heard of the sound. Being-heard is structurally phenomenal. . . . One can only divide this . . . by phenomenological reduction . . . [which] is indispensable to all analyses of being heard, whether they be inspired by linguistic, psychoanalytic, or other preoccupations.[88]

According to Derrida, the trace, or *différance*, is the 'being imprinted of the print'—*être imprimé de l'empreinte*—again, separable only through phenomenological reduction of materiality. This distinction allows for the articulation of difference (as such) and for consciousness. It both accounts for and exceeds (the logic of) metaphysics. For Derrida: 'the trace is not more ideal than real . . . it is anterior to the distinction.'[89] Imprint is irreducible (and this irreducibility is devised by Derrida through a radicalization of Saussure's sound-image) to 'either traditional philosophical analysis or to any analysis such as that of linguistics which presumes to supersede the originary transcendental thrust of philosophy'.[90] The graphic or phonic sign is marginal, what is important is the 'middle ground' we reach. Beardsworth explains: 'neither suspended in the transcendental nor rooted in the empirical, neither in philosophy nor in any empirical negotiation of the world that refuses to pass through the transcendental. The refuse to pass through the transcendental condemns one to a description of the fact of difference which is unable to take into proper account the necessity and economy of violence, its "genealogy". It thereby repeats the naïve violence particular to the oppositional axiomatic of metaphysics.'[91]

For Derrida, the generalization of writing is confirmed by the analogy with writing that Saussure makes precisely when bracketing the material. This also allows him to confirm that 'arche-writing, as an *originary structure of repetition*, constitutes the structure of the "instituted trace" which comprehends the foundation, exclusion and contradiction of (the history of) linguistics.' To quote Beardsworth again:

> [f]or a *t* to have identity as a *t*, it must be repeated. There can be no identity without repetition; and yet, this very repetition puts in question the identity which it procures, since repetition is always made in difference. Absolute repetition is impossible in its possibility, for there can be no repetition without difference. The concepts of repetition and difference form the precipitate of the metaphysical dissolution of an originary aporetic structure of repetition which Derrida calls 'arche-writing' or 'trace'.[92]

This passage clarifies how the structure of repetition is at the basis of the process of signification. The mark is subject to the law of repetition in difference: the opposition between speech and writing is, for Derrida, a determining one in metaphysics. The law of this repetition, and of the metaphysical decision to transform it into an opposition, can be traced through the linguistic mark. However, this characterizes not only the linguistic mark, but *all* marks: 'all marks are only possible within this process of idealization'.[93] In Beardsworth's words:

> 'Arche-writing' brings together, therefore, the analysis of originary violence specific to the elaboration of the trace with the simultaneous reinscription by Derrida of the opposition between the transcendental and the empirical. In other words, it brings together Derrida's analysis of the institution (here, of linguistics)

with his renegotiation of the frontiers between philosophy and the empirical sciences. Indeed, the one analysis cannot be separated from the other. The method of deconstruction constitutes from the beginning both a reinscription of the empirico-transcendental difference and an analysis of the irreducibility of violence in any mark.[94]

In sum, for Derrida we need to have a sense of writing in order to gain a sense of orality. 'Writing' then takes precedence over orality not because writing historically existed before language, but because we must have a sense of the permanence of a linguistic mark in order to recognize it and to identify it. Ultimately, the sense of writing is necessary for signification to take place. But we can have a sense of the permanence of a mark only if we have a sense of its inscription, of its being embodied in a material surface. In other words, although we recognize the written form of a grapheme (let's say 't') only by abstracting it from all the possible empirical forms a 't' can take in writing, nonetheless we need such an empirical inscription to make this recognition possible. This is what Derrida means when he says that 'the transcendental' is always impure, always already contaminated by 'the empirical'. In other words, language itself is material for Derrida; it needs materiality (better: it needs the possibility of 'inscription') to function as language.

This interpretation contrasts with the consolidated (Anglo-American) reception of 'poststructuralism' and of Derrida's thought as unaware of the material aspects of culture, society, economics, and politics (according to Derrida's famous statement that 'there is nothing outside the text'). On the contrary, what I want to emphasize here is that textuality and materiality are not opposed. There is no actual need—as it is often claimed—to 'go back to materiality' after the 'linguistic turn' in cultural studies, because materiality has always been there.[95] Writing is material because materiality is the condition for writing itself, and for signification.

If materiality is the condition for signification, then every code is material. More precisely: the condition for code to function is the possibility of inscription. This is not the same as saying that software always has material and semiotic characteristics, or that it involves physical apparatuses as well as information, microcircuits and Boolean logics, the social and the technical. Of course this is all true, but what a material understanding of software means in addition is that software can function only through materiality—not because it has to run on a processor, nor because there are economic forces behind it, but because, as every other code, it functions only through materiality, since materiality is what constitutes signs (and therefore codes). Moreover, writing is based on the very same possibility of material inscription, and the fact that it has been posited in a significant relation with software by software engineering should come as no surprise at this point. Paraphrasing Bruno Latour, it might be said that 'we have never been immaterial.'[96]

But if every code is material, and if the material structure of the mark is at work everywhere, how are we supposed to study software as a historically specific technology? Two questions resurface here—namely, Stiegler's question regarding the relationship between originary technicity and historically specific technologies (how is one supposed to distinguish software from other historically specific technologies in a way that is meaningful for understanding the relationship between technology and the human?) and Hayles's question on the relationship between code, writing, and language (to what extent and in what way can software be distinguished from other historically specific material inscriptions?).

As I have shown in Chapter 1, it is Stiegler' reworking of the concept of originary technicity that establishes the foundation for the concrete investigation of historically specific technologies. This is also where Stiegler breaks with Derrida's thought by substantially revising the concept of 'writing in general'. For Derrida 'writing in general' (or 'archewriting', or *grammé*) represents the possibility of making conceptual distinctions, and thus the moment that precedes all conceptual oppositions. This is why for Derrida 'writing in general' cannot be studied empirically and historically. In *Technics and Time*, Stiegler appropriates Derrida's understanding of writing in general in terms of the 'originary prostheticity of the human'. To present but a quick summary of Stiegler's argument, for him it is technology (or 'organized inorganic matter') that ultimately supports the possibility of material inscription, which in turn is constitutive of transcendence and of the possibility of thought. (For Stiegler, humanity 'transcends' its genetic program in pursuing life through means other than life). But, as we have seen in Chapter 1, organized inorganic matter changes in time. Therefore, for Stiegler it is possible to write a history of organized inorganic matter—that is, a history of the *grammé*, or of the conditions of thought. This is why Beardsworth observes that Stiegler 'pushes' deconstruction towards technology and that Stiegler's approach 'repeats and transforms' Derrida's strategy: '[i]t repeats the strategy in the sense that Derrida's philosophy constitutes a transformation of oppositional logic, notably that between the transcendental and the empirical, which works towards a more refined thinking of finitude. It transforms Derrida's relation to the tradition of philosophy, however, by giving in the aporetic terms of "matter" a history of what precedes oppositional logic.' In other words, for Stiegler, 'the irreducibility of the *grammé* must at the same time be articulated in terms of its (historical) differentiations in different "systems of writing."'[97]

I take Stiegler's rereading of Derrida as my starting point in order to answer the question of the specificity of software as a material inscription and of the relationship that software entertains with other material inscriptions such as those named by Hayles 'code', 'writing', and 'language'. I want to advance the following proposition: there is no general approach to software that can establish once and forever how software works and what place it occupies in relation to 'writing' and 'language'. As I have shown earlier on, software has never been univocally defined by any disciplinary field (including technical ones) and it takes different forms in different contexts. The definition of software that constitutes the conceptual foundation of software engineering is a particularly interesting starting point for the investigation of software because it establishes a very strong relationship between software and writing. The question regarding the specificity of software can thus be reformulated as such: how does 'software' emerge as a historically specific technology in the discourses and practices of software engineering and in what relationship with 'writing' and 'language'? Although this question will find an answer only in the following chapter, I want to suggest here that the specificity of software as it is conceptualized in the disciplinary field of software engineering resides precisely in the relationship it entertains with writing—or better, with a historically specific form of writing. It is in this relationship that software finds its singularity.

In Chapter 3, I will show how 'software', 'writing', and 'code' emerge together—and actually constitute each other as constantly shifting terms—in the context of software engineering at the beginning of the 1960s. However, what I want to emphasize here is that there is no 'writing' prior to 'software'—that is, there is no such writing as the one that emerges in and with software engineering. Such kind of writing emerges only there, and only in relation with software and code. One cannot exist without the others. Although this co-emergence constitutes the historical specificity of software engineering, it cannot be said that it also constitutes the specificity of 'all software', since the definition of software varies in time and space. In Chapter 3, I will also argue that there is actually no general

intermediation between software, writing, and code. Rather, a co-
constitution of software, writing, and code can be identified in and
as software engineering. The singularity of 'writing software' — as a
'singular' practice distinct from other kinds of 'singular' practices,
such as 'writing literature', or 'writing electronic literature' — is precisely this: that it emerges in relation to software and code in the
(again, 'singular') context of software engineering.

Even more importantly, one needs to be reminded that, in this
context, software transgresses the distinction between technics and
mnemotechnics established by Stiegler in *Technics and Time*.[98] Software — as it is defined by software engineering in the 1960s and
later — can be considered both as technics (since it makes things
'happen' in the world) and as mnemotechnics (since it is also a form
of recording). In Stiegler's terms, one could say that software is a
kind of mnemotechnics whose principal aim is *not* recording. However in Chapters 3 and 4, I will call into question this idea of the
'aim' of software and I will show how software (again, in the singular contexts that I will analyze) functions as a form of material inscription that continually *disrupts its aim* — that is, the aim for which
it is intended — by generating unforeseen consequences. Ultimately,
software as material inscription constantly undoes its own instrumentality — that is, what Derrida would name its 'complicity with
metaphysics' — by generating reinscriptions that cannot be fully anticipated by those interacting with software. In this sense software
can be understood as a *what* that shapes and is shaped by a *who*. The
unexpected consequences of software have to be dealt with by making decisions (for instance, by deciding if a certain unexpected behaviour is a malfunction or a welcome innovation). Thus, innovation
emerges as the articulation between the human and the technical.

As I will show in Chapter 4, open source emerges in the 1990s as
one of the unexpected consequences of the models of software engineering of the 1970s and 1980s. In fact, open source is the result of
an engineering effort to obtain usable software more rapidly and
efficiently than was allowed by previous models of software development. Of course, open source also has political implications,
since, for instance, it broadens the access of programmers and users
to certain stages of software development. However, the political
implications of open source emerge only in the context of this particular articulation of the human and the technical. From this point
of view, asking whether the unexpected consequences of software
are caused by the intrinsic dynamics of technology or by those of
society would not make sense. Instead, the social and the technical

co-emerge during the process of software inscription and—more broadly—of technological reinvention. As we shall see in Chapter 4, in open source, political practices co-emerge with technology.[99]

The political significance of a deconstructive analysis of software should be quite clear by now. However, in order to expand on this point briefly, let me turn for a moment to other examinations of the political implications of software studies, and particularly to Matthew Fuller's call for a critique of software in his 2003 book *Behind the Blip*. Fuller opposes the 'idealized', normative, functionalist description of software development that can be found in technical literature (which describes how to design software) to the more realistic approach of programmer's own accounts of their practice (which describe how software is actually designed). Furthermore, he points out how in technical literature software is viewed as instrumental ('a neutral tool'), while programmers' accounts offer an alternative to the instrumental conception of technology by taking into account 'other formations' than the technical (for instance, the cultural, the social, and the aesthetic). I can only suppose here that Fuller would view the technical literature of software engineering as an example of idealized description. And yet, my aim in Chapters 3 and 4 will be to demonstrate how technical literature can also be read in a non-functionalist and deconstructive way in order to unmask how the conceptual system of software works. Such a reading will also make apparent how the instrumental understanding of software in technical literature is much more controversial and unstable than Fuller assumes it to be. Of course, Fuller's strategy of opposing programmers' own accounts (how things 'really' are) to prescriptive technical literature (how things should be) can also be pursued. Nevertheless, I suggest that a complementary and equally interesting approach would involve showing how technical literature has been constituted as such—and thus, ultimately, how software has become what it is.

In other words, the deconstructive reading of software I am proposing in *Software Theory* is not the same as a critical reading of software in the sense propounded by Fuller. For Fuller, in order to produce some form of criticism of software, we need to focus on alternative software production and on those models of software (such as critical software, social software, and speculative software) that 'contain engines for its theorization'.[100] For instance, critical software is software that runs like commercial applications but has been 'fundamentally twisted to reveal underlying constructions' (of the user, the coding, and so on).[101] Social software (such as free software and open source) is accessible to people who are normally excluded from industrial software production. Finally, speculative software has a tight relationship with fiction and art, and it 'reinvents' and expands upon existing languages to 'explore the potentiality of all possible programming'.[102] In sum, Fuller suggests that the best critical approach to software is the production of alternative software that 'twists' or 'reveals' what is normally 'behind' software itself. Fuller's image of the 'blip' is of the utmost importance here. For him, 'behind the blip' we can find the social, economic, and political realm—and speculative software wants to 'intercept', 'map', and 'reconfigure' the social, the economic, and the political 'by means of the blips'.[103] According to Fuller, 'blips' are not signifiers 'but integral and material parts of events which manifest themselves digitally'. Speculative software is able to 'operate reflexively upon itself and the condition of being software' and makes visible the dynamics of the social and economical events it connects to.[104]

However interesting, Fuller's argument on the 'blip' remains partly problematic. While being aware that software is a social object, and remaining sympathetic with those approaches to software studies that aim at elucidating the political, social, and economic aspects of software production and consumption, I want to ask here whether the notion of 'making the social apparent' in software conveys an implicit dream of transparency. In other words, exactly what kind of demystification (if any) is at work in critical and speculative software? It seems to me that, to a certain extent, critical and speculative software always run the risk of substituting demystification with some other mystification—that is, the mystification of immediacy, or the mystification of demystification.

Once again, and in order to avoid some dead ends, an alternative approach would be to keep in mind that contemporary technology (including software) is already in deconstruction. But a deconstructive analysis of software does not aim at making apparent 'the social' behind and through software ('the blip'). It rather aims at dealing with the co-emergence of the social and the technical in software, understood within the framework of originary technicity. It also aims at engaging—on a political and philosophical level—with what remains unthinkable in software, with the unforeseen consequences of technology and with its capacity of bringing forth the unexpected.

NOTES

1. Timothy Clark, "Deconstruction and Technology," in *Deconstructions. A User's Guide*, ed. Nicholas Royle (Basingstoke: Palgrave, 2000), 248.

2. Clark, "Deconstruction and Technology," 247.

3. Clark, "Deconstruction and Technology," 247.

4. Jacques Derrida, "The Principle of Reason: The University in the Eyes of Its Pupils," *Diacritics* 13, no. 3 (1983): 14.

5. Jacques Derrida and Bernard Stiegler, *Echographies of Television: Filmed Interviews* (Cambridge: Polity Press, 2002), 45.

6. Derrida and Stiegler, *Echographies of Television*, 71.

7. Clark "Deconstruction and Technology," 249.

8. Jacques Derrida, *Specters of Marx: The State of the Debt, the Work of Mourning, and the New International* (New York and London: Routledge, 1994), 54.

9. Derrida, *Specters of Marx*, 169.

10. Clark "Deconstruction and Technology," 249.

11. On the romantic theory of genius, see Jacques Derrida, "Psyche: Inventions of the Other," in *Reading de Man Reading*, ed. Lindsay Waters and Wlad Godzich (Minneapolis: University of Minnesota Press, 1989), 58–60.

12. Derrida, *Specters of Marx*, 55.

13. Clark, "Deconstruction and Technology," 251.

14. Clark, "Deconstruction and Technology," 251.

15. As I will show later on in this chapter, Derrida and Stiegler hold different views on the emergence of technical innovation and on the possibility of an empirical investigation of originary technicity. In *Ecographies of Television* this in turn leads them to different perspectives on the possibility of determining the specificity of tele-technologies.

16. Clark, "Deconstruction and Technology," 252.

17. See, for example, Wendy Hui Kyong Chun, *Programmed Visions: Software and Memory* (Cambridge, MA and London: MIT Press, 2011); David Berry, *The Philosophy of Software: Code and Mediation in the Digital Age* (Basingstoke: Palgrave Macmillan, 2011); Florian Cramer, *Anti-Media: Ephemera on Speculative Arts* (Rotterdam: NAi Publishers and Institute of Network Cultures, 2013).

18. Katherine N. Hayles, *My Mother Was a Computer: Digital Subjects and Literary Texts* (Chicago: University of Chicago Press, 2005), 17–18.

19. Hayles, *My Mother Was a Computer*, 17.

20. Hayles, *My Mother Was a Computer*, 17; Jacques Derrida, *Positions* (London and New York: Continuum, 2004), 19.

21. Hayles, *My Mother Was a Computer*, 22.

22. Hayles, *My Mother Was a Computer*, 15.

23. Hayles, *My Mother Was a Computer*, 16.

24. Hayles, *My Mother Was a Computer*, 16.

25. Ferdinand de Saussure, *Course in General Linguistics* (Peru, IL: Open Course Publishing, 1988); Jacques Derrida, *Of Grammatology* (Baltimore: The Johns Hopkins University Press, 1976); Stephen A. Wolfram, *A New Kind of Science* (New York: Wolfram Media, 2002); Harold J. Morowitz, *The Emergence of Everything: How the World Became Complex* (Oxford and New York: Oxford University Press, 2002); Ellen Ullman, *Close to the Machine: Technophilia and Its Discontents* (San Francisco: City Lights Books, 1997); Matthew Fuller, *Behind the Blip: Essays on the Culture of Software* (New York: Autonomedia, 2003); Matthew G. Kirschenbaum, "Materiality and Matter and Stuff: What Electronic Texts Are Made Of," *Electronic Book Review* 12 (2002), http://www.altx.com/ebr/riposte/

rip12/rip12kir.htm , Bruce Eckel, *Thinking in C++* (Englewood Cliffs, NJ: Prentice Hall, 1995).

26. Hayles's concept of intermediation draws on what Jay Bolter and Richard Grusin have called 'remediation'—that is, 'the formal logic by which new media technologies refashion prior media forms' (Jay David Bolter and Richard Grusin, *Remediation: Understanding New Media* [Cambridge, MA: MIT Press, 2002], 273). With the more comprehensive term 'intermediation' she wants to emphasize the multiple causalities that influence interactions among media.

27. Wolfram, *A New Kind of Science*; Morowitz, *The Emergence of Everything*.

28. Hayles, *My Mother Was a Computer*, 30.

29. Hayles, *My Mother Was a Computer*, 22–23.

30. Hayles, *My Mother Was a Computer*, 23.

31. Hayles, *My Mother Was a Computer*, 20. For instance, Wolfram's work shifts from regarding computation as a way to conceptualize complex systems (as in the simulation of life processes by means of cellular automata running in a computer) to thinking computation as a process that 'actually generates reality' (Hayles, *My Mother Was a Computer*, 19).

32. Hayles, *My Mother Was a Computer*, 21.

33. Hayles, *My Mother Was a Computer*, 21–22.

34. Richard Doyle, *Wetwares: Experiments in Postvital Living* (Minneapolis: University of Minnesota Press, 2003); Hayles, *My Mother Was a Computer*, 22.

35. Hayles, *My Mother Was a Computer*, 41.

36. Hayles, *My Mother Was a Computer*, 41.

37. de Saussure, *Course in General Linguistics*, 25–26.

38. Hayles, *My Mother Was a Computer*, 42.

39. Jonathan Culler, *Ferdinand de Saussure* (New York: Cornell University Press, 1986), 33.

40. Hayles, *My Mother Was a Computer*, 43. A thorough discussion of the role played by material constraints in programming (for instance, in terms of time and memory resources) is provided by Jay D. Bolter, *Turing's Man: Western Culture in the Computer Age* (London: Duckworth, 1984).

41. Hayles, *My Mother Was a Computer*, 45; Friedrich Kittler, "There Is No Software," *CTheory*, October 1995, http://www.ctheory.net/articles.aspx?id=74.

42. Hayles, *My Mother Was a Computer*, 45.

43. Hayles, *My Mother Was a Computer*, 45.

44. Noam Chomsky, *Aspects of the Theory of Syntax* (Cambridge, MA: MIT Press, 1965). See also Andrew. S. Tanenbaum, *Structured Computer Organization* (Englewood Cliffs, NJ: Prentice-Hall, 1999).

45. Alexander Galloway, *Protocol: How Control Exists after Decentralization* (Cambridge, MA: MIT Press, 2004), 165.

46. John L. Austin, *How to Do Things with Words* (Oxford: Clarendon Press, 1972).

47. Austin, *How to Do Things with Words*, 112.

48. Hayles, *My Mother Was a Computer*, 49.

49. Hayles, *My Mother Was a Computer*, 50.

50. Hayles, *My Mother Was a Computer*, 50.

51. Katherine N. Hayles, *How We Became Posthuman: Virtual Bodies in Cybernetics, Literature and Informatics* (Chicago: University of Chicago Press, 1999), 274.

52. See, for instance, Karen Barad, "Posthumanist Performativity: Toward an Understanding of How Matter Comes to Matter," *Signs: Journal of Women in Culture and Society* 28, no. 3 (2003): 801–31; Eve Kosofsky Sedgwick, *Touching*

Feeling: Affect, Pedagogy, Performativity (Durham, NC and London: Duke University Press, 2003).

53. Judith Butler, *Excitable Speech: A Politics of the Performative* (New York and London: Routledge, 1997), 4.

54. Jacques Derrida, *Limited Inc.* (Evanston, IL: Northwestern University Press, 1988).

55. Derrida, *Limited Inc.*, 3.

56. Derrida, *Limited Inc.*, 4.

57. Derrida, *Limited Inc.*, 7.

58. Derrida, *Limited Inc.*, 12.

59. Derrida, *Limited Inc.*, 12.

60. Derrida, *Limited Inc.*, 13.

61. Derrida, *Limited Inc.*, 15.

62. Derrida, *Limited Inc.*, 16–17.

63. Hayles, *My Mother Was a Computer*, 56.

64. Hayles, *My Mother Was a Computer*, 56.

65. Hayles, *My Mother Was a Computer*, 60.

66. Wendy Hui Kyong Chun, *Control and Freedom: Power and Paranoia in the Age of Fiber Optics* (Cambridge, MA and London: MIT Press, 2006). See also Wendy H. K. Chun, *Programmed Visions: Software and Memory* (Cambridge, MA and London: MIT Press, 2011).

67. Hayles, *My Mother Was a Computer*, 61.

68. Hayles, *How We Became Posthuman*, 4–6.

69. Richard Beardsworth, *Derrida and the Political* (New York: Routledge, 1996), 7.

70. Beardsworth, *Derrida and the Political*, 7.

71. Beardsworth, *Derrida and the Political*, 8.

72. Beardsworth, *Derrida and the Political*, 8.

73. de Saussure, *Course in General Linguistics*, 15.

74. Beardsworth, *Derrida and the Political*, 9.

75. Jacques Derrida, *Of Grammatology*, 61.

76. Beardsworth, *Derrida and the Political*, 9.

77. de Saussure, *Course in General Linguistics*, 158.

78. Beardsworth, *Derrida and the Political*, 12.

79. Beardsworth, *Derrida and the Political*, 12.

80. As we have seen in Chapter 1, this point has been developed by Gary Hall in his analysis of cultural studies as a field particularly attentive to the institutional forces that shape academic knowledge, which in turn should pursue a tighter connection with deconstruction in order to strengthen this awareness. Cf. Gary Hall, *Culture in Bits: The Monstrous Future of Theory* (London and New York: Continuum, 2002).

81. Beardsworth, *Derrida and the Political*, 13.

82. Beardsworth, *Derrida and the Political*, 14.

83. de Saussure, *Course in General Linguistics*, 165–66.

84. Beardsworth, *Derrida and the Political*, 15.

85. Beardsworth points out how Derrida radicalized Husserl in his introduction to *The Origin of Geometry* and in *Speech and Phenomena*. For this reason, according to Beardsworth, Derrida is so attentive to Saussure's 'reduction of the phonic substance of the sign' (Beardsworth, *Derrida and the Political*, 15). Beardsworth also highlights how the importance of Derrida's understanding of the reduction of phonic substance has been generally underestimated.

86. Beardsworth, *Derrida and the Political*, 16.

87. Beardsworth, *Derrida and the Political*, 16.

88. Derrida, *Of Grammatology*, 93.

89. Derrida, *Of Grammatology*, 95.

90. Beardsworth, *Derrida and the Political*, 17.

91. Beardsworth, *Derrida and the Political*, 17.

92. Beardsworth, *Derrida and the Political*, 17.

93. Beardsworth, *Derrida and the Political*, 18.

94. Beardsworth, *Derrida and the Political*, 18.

95. For a discussion of textuality/materiality as one of the 'false oppositions' of new media studies, see Sarah Kember and Joanna Zylinska, *Life after New Media: Mediation as a Vital Process* (Cambridge, MA and London: MIT Press, 2012).

96. Bruno Latour, *We Have Never Been Modern* (Cambridge, MA: Harvard University Press, 1993).

97. Richard Beardsworth, "From a Genealogy of Matter to a Politics of Memory: Stiegler's Thinking of Technics," *Tekhnema: Journal of Philosophy and Technology* 2 (1995): 92.

98. Bernard Stiegler, *Technics and Time, 3: Cinematic Time and the Question of Malaise* (Stanford, CA: Stanford University Press, 2011), 131.

99. Beardsworth points out how, in Derrida's version of deconstruction, politics is based on the promise of a future that must always be thought as radically 'other', while for Stiegler the future of politics coincides with the reinvention of the relationship between the human and the technical. (Here one is reminded also of the divergence between Derrida and Stiegler on how to think innovation in *Echographies of Television*). For this reason, Stiegler has been sometimes criticized for overdetermining the concept of the future (see, for instance, Mark Hansen, "'Realtime Synthesis' and the Différance of the Body: Technocultural Studies in the Wake of Deconstruction," *Culture Machine* 5, 2003, http://www.culturemachine.net/index.php/cm/article/view/9/8). My analysis is meant to show how 'otherness' can emerge from the interaction between technology and the human as what remains 'unprogrammable' in software. Beardsworth's argument substantially supports mine, since he remarks how Stiegler's politics of memory—which 'is nothing but the struggle to remember, actively, the relation between the technical and the human'—can be articulated in terms of 'multiple, heterogeneous sites of invention' (Richard Beardsworth, "From a Genealogy of Matter to a Politics of Memory," 115).

100. Fuller, *Behind the Blip*, 22.

101. Fuller, *Behind the Blip*, 25.

102. Fuller, *Behind the Blip*, 30.

103. Fuller, *Behind the Blip*, 31.

104. Fuller, *Behind the Blip*, 31–32.

THREE

Software as Material Inscription

The Beginnings of Software Engineering

In the beginning was the word, all right—[general laughter] but it wasn't a fixed number of bits!

—Peter Naur and Brian Randell

In an article published in 2004 in the *Annals of the History of Computing*, Michael S. Mahoney writes:

> Dating from the first international conference on the topic in October 1968, software engineering just turned thirty-five. It has all the hallmarks of an established discipline: societies (or subsocieties), journals, textbooks and curricula, even research institutes. It would seem ready to have a history. Yet, a closer look at the field raises the question of just what the subject of the history would be. [1]

Mahoney points out that, although it is not hard to find definitions of software engineering throughout technical literature—for instance, in his 1989 paper entitled 'The Software Engineering Process', Watts Humphrey, a leading practitioner in the field, defines it as 'the disciplined application of engineering, scientific, and mathematical principles and methods to the economical production of quality software'—it is also rather easy to come across doubts as to whether software engineering's current practice meets such criteria. [2] For example, in an article published at about the same time, Mary Shaw—herself a distinguished scholar and practitioner in the field—muses whether software engineering is an engineering discipline at all, and states: 'Software engineering is not yet a true engineering discipline, but it has the potential to become one.'[3] 'From the outset', Mahoney comments, 'software engineering conferences have routinely begun with a keynote address that asks "Are we there yet?" and proposes yet another specification of just where "where" might be.'[4]

Being interested in writing a history of software engineering, Mahoney is particularly troubled by the fact that 'the field has been a moving target for its own practitioners' from its very beginning, and that practitioners openly disagree on what it is. Historians—he suggests—can as readily write a history of software engineering 'as the continuing effort of various groups of people engaged in the production of software to establish their practice as an engineering discipline'—that is, a history of this process of self-definition.[5] Such an approach would immediately pose a number of questions, such as which model of engineering software practitioners refer to. For instance, an oft-quoted passage from the introduction to the proceedings of the first NATO Conference on Software Engineering, held in Garmisch in 1968, declares: '[t]he phrase "software engineering" was deliberately chosen as being provocative, in implying the need for software manufacture to be based on the types of theoretical foundations and practical disciplines that are traditional in the established branches of engineering.'[6]

The definition of software engineering given in the proceedings of the Garmisch Conference is provocative indeed, since it leaves all the crucial terms undefined. For instance, it is unclear what 'manufacturing' software means, or what the 'theoretical foundations and practical disciplines' that underpin the 'established branches of engineering' are. Furthermore, the term 'traditional' hints at a search for historical precedents—and yet the above passage seems to leave open the question of how the existing branches of engineering have taken their present form. On the other hand, the definition of software engineering also concerns its 'agenda'—that is, what practitioners agree ought to be done, 'a consensus concerning the problems of the field, their order of importance or priority, the means of solving them (the tools of the trade), and perhaps most importantly, what constitutes a solution'.[7] Much of the disagreement among the participants in the first NATO Conference on Software Engineering rested both on their different professional backgrounds and on the conflicting agendas they brought to the gathering. None of them— as Mahoney emphasizes—was a software engineer, since the field did not exist. Rather, they came from quite varied professional and disciplinary traditions.

Albeit Mahoney presents self-reflexivity as an inconvenience, the fact of being a 'moving target' makes software engineering particularly valuable as a starting point for my investigation of software. In Chapter 2, I suggested that software's specificity as a technology resides precisely in the relationship it entertains with writing, or, to be more precise, with historically specific forms of writing. In this chapter, I want to illustrate the coemergence of software and writing—or, even better, of 'software', 'writing', and 'code' as constantly shifting terms which are actually constitutive of one another—in the discourses and practices of software engineering in the late 1960s, at the beginnings of the discipline. The aim of this chapter is not to produce a historical overview of the field in Mahoney's sense. Instead, the aim is to investigate the way in which software engineering has been instituted as a field of knowledge—a process that involved the establishment of its own object ('software'), itself involved in a mutually constitutive relationship with two other entities ('writing' and 'code'). In order to do this, in this chapter, I offer a deconstructive reading of the foundational text of software engineering—namely, the report of the first Conference on Software Engineering, convened by the NATO Science Committee in 1968 in Garmisch (Germany)—the first-ever conference on software engineering.[8] A second conference was held in 1968 in Rome (Italy). However, one year later things had changed significantly in software engineering and the climate of the Rome conference was far less enthusiastic than that of the first one. The main point of interest in the Rome report is a growing awareness of the lack of communication between software practitioners and the academic world—something that did not seem to affect the participants of the Garmisch conference. Furthermore, all the essential issues of software engineering are set out in the Garmisch conference report.[9]

The Garmisch report is an extremely interesting text. It was produced as a reworking of conference materials plus a selective transcription of the discussion that took place at the conference. The discussion was attended by several reporters taking notes and partially recorded on magnetic tape. The recording was then correlated with the reporters' notes and selectively transcribed. Parts of the discussion were lost in the process, which worked as a filter. Ultimately, the report constitutes a narrative based on a collection of direct quotations, which were meant to represent the participants' conflicting point of views, sparsely interpolated by the editors' com-

ments and followed by a selection of working papers contributed by
the participants and included as appendices. This widely read text
fundamentally shaped the field of software engineering in the following years, and it did so through the selection, inclusion, and
exclusion of problems and topics. Following Gary Hall, it could be
said that the report constitutes the first attempt at building an
archive for software engineering and at performing software engineering as a discipline.[10]

Historically, software engineering emerged from a crisis, the so-
called 'software crisis' of the late 1960s. As Brian Randell—editor of
the reports of the 1968 and 1969 conferences—recalls later on in his
article 'Software Engineering in 1968', it was 'the Garmisch conference that started the software engineering bandwagon rolling'.[11]
According to Randell, one of the most significant aspects of the
Garmisch conference was the willingness of the participants to admit 'the extent and seriousness of current software problems' of the
time.[12] For instance, during the conference Edsger W. Dijkstra re-
portedly stated that '[t]he general admission of the existence of the
software failure in this group of responsible people is the most re-
freshing experience I have had in a number of years, because the
admission of shortcomings is the primary condition for improvement.'[13] Terms such as 'software crisis' and 'software failure' were
largely used during the Garmisch conference, and for this reason
many of the participants viewed the conference as a turning point in
their way of approaching software. In fact, with the Garmisch conference, software began to be conceptualized *as a problem*—and the
'software crisis' was constituted as a point of origin for the discipline of software engineering. From the very beginning, the participants in the Garmisch conference acknowledged that they were
dealing with 'a problem crucial to the use of computers, viz. the so-
called software, or programs, developed to control their action'.[14]
The very first lines of the Garmisch report establish a clear relationship between software and control, while at the same time charac-
terizing this relationship, as well as software itself, as problematic.
But why was software 'problematic' in the late 1960s?

The issues that the conference attempted to address were mainly
related to 'large' or 'very large' software systems—that is, systems
of a certain complexity whose development required a conspicuous
effort in terms of time, money, and the number of programmers
involved. As Randell comments in his recollections about the Garmisch conference (thus explaining NATO's interest in software engineering), 'it was the US military-industrial complex that first

started to try and develop very large software systems involving man-millennia of effort.'[15] Randell also mentions a paper presented by Joseph C. R. Licklider in 1969 as a contribution to the public debate around the Anti-Ballistic Missile (ABM) System (a complex project which contemplated the development of enormously sophisticated software) and eloquently titled 'Understimates and Overexpectations'. In his paper, Licklider provides a vivid picture of the gap between the military's goals and its achievements. He states: '[a]t one time, at least two or three dozen complex electronic systems for command, control and/or intelligence operations were being planned or developed by the military. Most were never completed. None was completed on time or within the budget.'[16] Even more importantly, Randell adds the following comment:

> I still remember the ABM debate vividly, and my horror and incredulity that some computer people really believed that one could depend on massively complex hardware and software systems to detonate one or more H-bombs at exactly the right time and place over New York City to destroy just the incoming missiles, rather than the city or its inhabitants.[17]

Quite obviously, Randell's horror at the excessive self-confidence of some software professionals stems from the connotative association between technology, catastrophe, and death in a cold-war scenario. As we shall see in a moment, 'horror'—a powerful emotion—is the result of the anticipation of the consequences of technology combined with the awareness of its intrinsic fallibility.

However, by the late 1960s large-scale systems were not unique to the military scene. For instance, computer manufacturers had started to develop operating systems.[18] The late 1960s operating systems were much more complicated than their predecessors. For instance, release 16 of OS/360 was announced in July 1968 and contained almost one million instructions.[19] Specialized real-time systems were also being developed, such as the first large-scale airline-reservation system, the American Airlines SABRE system.[20] The costs incurred in developing them were immense, and they were very much in the public's eye. Moreover, some of these systems (such as TSS/360, an alternative to the operating system OS/360) kept performing poorly notwithstanding the vast amount of resources lavished on them by their manufacturers—and the professionals involved in these projects felt the pressure of the public opinion. Hal Helms is reported to have presented these dramatic figures at the Garmisch conference: ten thousand installed computers in Europe alone, a number increasing 'at a rate of anywhere from 25 percent to 50 percent per year'; furthermore, software development would soon involve 'more than a quarter of a million analysts and programmers'.[21] The situation is not only measured in terms of the number of software professionals involved, but also of cost and effort. During the conference E. E. David pointed out that T. J. Watson—IBM's founder—had estimated the cost of OS/360 development at over $50 million a year, and at least five thousand man-years, while TSS/360 was probably in the one hundred man-years category. The speed of software growth, according to the editors, was perceived by the conference participants with more 'alarm than pride.'[22]

One must be reminded at this point of the importance of the relationship between contemporary technology and speed. In Bernard Stiegler's words, it is the contemporary 'conjunction between the question of technics and the question of time' that 'calls for a new consideration of technicity'.[23] Stiegler hints to the 'dis-adjustment' between society and technology due to the speed of the latter. In a sense, the Garmisch report is concerned precisely with this problem. In fact, the Garmisch conference report was produced expressly to serve as an instrument for managers of the private and public sectors and policy makers to anticipate and evaluate the consequences of fast-developing technology. The participants in the Garmisch conference viewed society at large as mainly concerned with the problem of the reliability of software and with its costs, and they measured the relationship between software and society in terms of 'impact'. However, it is precisely this opposition between society and technology that seems not to hold everywhere in the Garmisch report. For instance, participant E. E. David describes the process of software growth in the following terms:

> In computing, the research, development, and production phases are often telescoped into one process. In the competitive rush to make available the latest techniques, such as on-line consoles served by time-shared computers, we strive to take great forward leaps across gulfs of unknown width and depth. In the cold light of day, we know that a step-by-step approach separating research and development from production is less risky and more likely to be successful. Experience indicates that for software tasks . . . estimates are accurate to within 10-30 percent in many cases. This situation is familiar in all fields lacking a firm theoretical base. Thus, there are good reasons why software tasks that

include novel concepts involve not only uncalculated but uncalculable risks.[24]

David focuses here on the pace of software growth. The competition between computer manufacturers forces software professionals to confuse ('telescope') research and production, which should remain separate. Therefore, the uncertainties typical of research (here intended as the development of innovative software) spread to production. David's metaphor opposes 'leaps' to 'steps'. The leap is for him a dangerous way to move forward, motivated by the lack of knowledge. The step-by-step approach would be a safer way—not, it seems, to slow down the growth of software, but to make the speed of such growth more manageable. As we have seen, the participants in the Garmisch conference had to face some major doubts concerning large-scale software systems: were such systems actually feasible? In David's terms, the question could have been reformulated as follows: was the speed of software growth actually manageable? Importantly, David attributes the necessity of taking big leaps forward to the lack of a 'firm theoretical basis': in other words, the inability to estimate the feasibility of a software project in a reliable way leads to the impossibility of carrying it out step by step, and ultimately to its failure. The failure of a software project then seems to be related to the failure of the management of time.

According to David, software professionals are fundamentally concerned not just with risk—that is, with the possibility of failure—but also with 'uncalculated' and 'uncalculable' risks. It seems quite understandable that certain risks cannot be calculated due to the lack of accurate knowledge. What is really surprising is David's use of the expression 'uncalculable'. It is not quite common for software professionals and engineers to acknowledge that a technical project involves uncalculable risks. Although the participants in the Garmisch conference must not have been aware of this, the concept of calculability of time has a distinct Heideggerian echo. For David, the concept of risk and calculability are both related to the future: estimates are the expression of a calculability of the future, they actually *presuppose* the calculability of the future. However, it is precisely this faith in the calculability of time, and therefore in the feasibility of software projects, that is put into question in the Garmisch report and in its narrative of the 'software crisis' as the source of technological 'horror'. But to what extent can the uncalculability lamented by David be linked to the 'unforeseen consequences' that for Derrida are always implicit in contemporary technology, thus making it 'technology-in-deconstruction'?[25] In what way did the participants in the NATO conferences explain the 'uncalculability' of technology?

The Garmisch conference report is dominated by a widespread recognition that 99 percent of software systems work—as Jeffrey R. Buxton states—'tolerably satisfactorily'.[26] Only certain areas are viewed with concern. Kenneth W. Kolence comments:

> The basic problem is that certain classes of systems are placing demands on us [software professionals] which are beyond our capabilities and our theories and methods of design and production at this time. There are many areas where there is no such thing as a crisis—sort routines, payroll applications, for example. It is large systems that are encountering great difficulties. We should not expect the production of such systems to be easy.[27]

We already know that the risky 'classes' of systems are large-scale and real-time ones. Nevertheless, Kolence seems to take the argument a step further and to relate the uncalculability of software development to certain demands posed by society that go beyond the technological capabilities of the time. In other words, not only do the conference participants feel the pressure of social demands on them; they also feel that software development reaches its point of crisis when society pushes the boundaries of state-of-the-art technology. But do these demands come from society or from technology itself? This question is at work in the whole of the Garmisch report and it silently destabilizes the separation between the technical and the social. Actually, it is precisely when dealing with the issue of the responsibility for the technological risk that the conference participants seem to be confronted with the impossibility of separating technology from society. For instance, Ascher Opler states:

> I am concerned about the current growth of systems, and what I expect is probably an exponential growth of errors. Should we have systems of this size and complexity? Is it the manufacturer's fault for producing them or the users' for demanding them? One shouldn't ask for large systems and then complain about their largeness.[28]

Opler's observation is intriguingly ambiguous. He asks whether the responsibility for the rate of the growth of technology must be attributed to the users or to the producers of technology. The undecidability of this dilemma leaves its mark on the field of software engineering and especially on its relationship with technological failure and ultimately with the management of time. On the one hand, the participants in the Garmisch conference seem to acknowledge that risks are implicit in software, and that software fallibility is unavoidable. This is what David and Fraser state: '[p]articularly alarming is the seemingly unavoidable fallibility of large software, since a malfunction in an advanced hardware-software system can be a matter of life and death.'[29] On the other hand, it is claimed that risks could be avoided if an appropriate and effective 'theory' of the development of software was to be produced. From this second point of view, the approach to software development must be 'systematic', and therefore it must become a form of engineering.[30] However, these two points of view are entangled, and one does not exist without the other. As a result, Stanley Gill concludes: '[i]t is of the utmost importance that all those responsible for large projects involving computers should take care to avoid making demands on software that go far beyond the present state of technology unless the very considerable risks involved can be tolerated.'[31]

This quotation might sound like an attempt to discharge the responsibility for technological risk on society. In fact, it requires deeper analysis, since in what way could policy makers evaluate risks that they do not know? Software professionals are the ones who are expected to have such knowledge. An Habermasian answer might suggest that policy makers should be better informed of technological risks and able to discuss them freely.[32] But what Gill is actually saying here is that society shall not make demands that can be met only by exceeding the current state of technology. Here we are confronted with one of the 'points of opacity' of the foundational narrative of software engineering.[33] Indeed, it seems to me that the irreconcilability of these two aspects—and therefore the necessity of calculating incalculable risks, and of attributing responsibility for them—is a point where software engineering 'undoes itself' precisely at the moment of its constitution. What Gill means here is that society needs to take responsibility for an incalculable risk. The real problem here is the incalculability of the speed of technological growth—that is, of the rate at which the state of technology is exceeded.

In sum, at the end of the 1960s, software engineering as a discipline with a theoretical foundation is called for in order to avoid the (unavoidable) fallibility of technology—a fallibility that constitutes the risk posed by technology, or, better, technology *as* a risk. This 'point of opacity' suggests that software engineering establishes itself *as a theory* of technology by expelling fallibility from technology—but such a fallibility (the unexpected consequences of technology) is perceived as intrinsic to technology itself, and it is exactly what allows software engineering to exist (that is, the reason why software engineering is called for). In other words, software engineering performs an impossible expulsion of constitutive failure from technology, with this move establishing itself as a discipline. Since such an expulsion is performed through the calculation of time, it can also be said that in software engineering the calculability of time is undone in its very constitution. Furthermore, the Garmisch report offers an interesting perspective on the contemporary dis-adjustment between social and technical systems: society is instituted in the report as that which places risky demands on technology—while at the same time the report declares technology as constitutively fallible, as something that intrinsically incorporates unforeseen consequences.[34] Therefore, the projection of the fallibility on society—that is, on the demands that society poses to technology—is the way in which the conference participants both assume and discharge responsibility for the technological risk: they cannot actually maintain the boundary between technology and society, because this boundary keeps becoming undone. In Heideggerian terms, Randell's 'horror' is the result of anticipation *plus* the fallibility of technology. In a way, it can be said that, contrary to Heidegger's understanding of the relationship with death as constitutive of a temporality which is more 'authentic' than the temporality of calculation, in software engineering the question of death (for instance, the death of New York's inhabitants caused by a ballistic device gone wrong) is dealt with *as* a problem of calculation.

In order to understand how the calculation of time is performed in software engineering, let me now examine what is meant by the 'systematic approach' to software that the participants in the Garmisch conference recommend. As we have seen earlier on, the term 'software engineering' hints at the need for a theoretical foundation of software development. Throughout the report, the editors and the conference participants point out how the discipline is 'in a very rudimentary stage of development as compared with the established branches of engineering', and especially with hardware production.[35] For instance, Doug McIlroy observes: '[w]e undoubtedly get the short end of the stick in confrontation with hardware people because they are the industrialist and we are the crofters.'[36] By 1968, the fact that 'software was an important commodity in its own right' was widely recognized.[37] Just in the United States there were around five hundred organizations concerned with selling and/or producing software—albeit the term 'software' had made its appearance in normal business parlance only a few years earlier.[38] Nevertheless, the process of developing software was not well understood yet. For this reason, McIlroy presented a paper on software production, entitled '"Mass-Produced" Software Components', at the Garmisch conference. Published among the appendices to the conference report, this paper investigates 'the prospects for mass-production techniques in software' and recommends the creation of software components according to the same criteria that regulate the production of hardware components. For McIlroy, one important reason for the weakness of the software industry in the late 1960s was the absence of a software components subindustry. With formidable insight he wrote:

> I would like to see components become a dignified branch of software engineering. I would like to see standard catalogues of routines, classified by precision, robustness, time-space performance, size limits, and binding time of parameters. . . . I want to have confidence in the quality of the routines. . . . What I have just asked for is simply industrialism, with programming terms substituted for the more mechanically oriented terms appropriate to mass production. I think that there are considerable areas of software ready, if not overdue, for this approach.[39]

McIlroy approaches software as a product and software production as a formalizable—and to some degree automatizable—process. His paper makes obvious how in the late 1960s the concept of engineering was introduced into the software field in connection with the view of software as a commodity and an industrial product. The standardized production of software could not be accomplished unless the process of software development was well understood, and the Garmisch conference focused precisely on pursuing such an understanding.

In the Garmisch report software production is depicted as an unfeasibly expensive trial-and-error process. For instance, Ronald Graham of Bell Labs maintains that '[t]oday we tend to go on for years, with tremendous investments to find that the system, which was not well understood to start with, does not work as anticipated. We build systems like the Wright brothers built airplanes—build the whole thing, push it off the cliff, let it crash, and start over again.'[40] The wording of Graham's passage is particularly interesting because it displays a certain confusion between the conceptualization of a software system and the understanding of its development. Such confusion returns over and over again throughout the Garmisch conference report. While a software system is mainly envisaged as a set of interrelated components that work together in order to achieve some objective, its development is understood as the process through which such a system is constructed — or better, as the conference participants commonly say, 'written'. Although the report focuses on understanding and defining the process of software development, the system and its development are never clearly separated. For instance, in Graham's passage the failures of the process of development (in which investment and time spin out of control) are caused by the poor understanding of the system in the first place. I want to argue here that the distinction between the software system and its development — that is, between process and product — is another 'point of opacity' of software engineering since, albeit necessary, it cannot be kept up at all times. Even more difficult is to measure the progress of a software project. In fact, as Fraser notices, 'program construction is not always a simple progression in which each act of assembly represents a distinct forward step and . . . the final product can be described simply as the sum of many sub-assemblies.'[41]

Fraser depicts software development as a complex and inordinate process, which software professionals need to understand in order to make it more linear. Or even better, they must acknowledge that such a process is not in fact linear, but at the same time they have to avoid backward steps at all cost. An ambivalence can be detected here in the conceptualization of software growth, which is considered a risk (since it implies failures) but at the same time a proof of success (and therefore it must be maintained by avoiding backward steps). Quite clearly, even more than the pressure of society, it is the implicit fallibility of software that worries conference participants. Such a fallibility needs to be dealt with through the steady movement towards an objective, thus calling for a linearization of time.[42] In turn, this appeal to the linearization of time is part of the general attempt to 'control' software — that is, to think software in instrumental terms.

Accordingly, in this first section of the report the focus shifts gradually from 'engineering' as a disciplinary model towards its specific approach to time: engineering involves planning — that is, the calculation of time — and systematic thinking. This view is supported by comments offered two decades later by numerous scholars. For instance, in her 1989 article entitled 'Remembrances of a Graduate Student', Mary Shaw recalls 1968 as a 'memorable year' and explains: 'it was the year the software research community really started thinking systematically, even formally, about software structures.'[43] She claims that the Garmisch conference represented a turning point in the passage from ad hoc to systematic practice and that it 'played a major role in shifting our thinking from *programs* to *software*'.[44] Thus, the term 'software' itself becomes the carrier of the idea of systematicity and of the calculation of time.

In sum, software engineering makes its appearance in the context of the early industrialization of software production and in opposition to the concept of craftsmanship, and it relies on a model of engineering as the calculation of time. The main object of software engineering as a discipline is the calculation of the time of software development in order to industrialize it. This calculation is to be realized through the formalization of the process of software development, and the Garmisch conference report contains a number of graphical and verbal models of software development that were proposed at the conference in order to attempt such formalization, such as the one offered by Calvin Selig.[45] With little variation, these schemes became the typical representations of what would later be called the 'software life cycle'. In the software life cycle, the process of software production is represented as the sequence of different stages, where each stage can be described as a period of intellectual activity which produces a usable product that constitutes the point of departure for the subsequent stage. At the beginning, a 'problem' is identified to which the software system will be the answer. At this point in the report, it must be noted, software has shifted from being a problem to being a solution, while the problem has been relocated to the external world. This again will later become part of the methodological core of software engineering: there are problems *out there* that software helps to solve.

The first stage of software development, named 'analysis', produces a description of the problem—which is substantially a document written in a natural language. The stage of analysis is variously enmeshed with the following phase of design (the two terms are used quite interchangeably in the report), which constitutes a refinement of the problem description in order to propose a software system that solves it. The phase of design produces a complete system specification—that is, a stable description of what the software system is supposed to do. The system specification is also the point of departure for the stage of implementation, which basically determines how the system will do what it is supposed to do. However, in the Garmisch report the boundaries between these stages are quite blurred and the terminology referring to the different stages of software development is remarkably unstable. The only clear separation is between the 'what' and the 'how': one must define 'what' a system is supposed to do before developing it, and the development of the system is the determination of 'how' the system does what it is supposed to do. Yet, even this distinction is put into question later on in the report, and the boundary between the 'what' and the 'how'—as I will show in a moment—turns out to be quite difficult to maintain. However, the stage of implementation 'translates'—a term attributed to Selig—the specification of the system into the 'actual' system (the 'working system')—that is, a software system that can be installed on a computer. Before it is delivered to the final user, the system needs to be tested—namely, it must be verified whether the system actually does 'what' it is meant to do and whether it does it 'how' it is meant to do it.[46] This stage is variously called 'testing', 'acceptance', or it can be incorporated into a general stage of 'maintenance', which comprises all the activities performed on a completely operational software system, such as corrections and modifications that might be carried out even after the system has been delivered to its users. As soon as it becomes operational, the system starts to become obsolete—and it will finally be abandoned in favour of newer systems.

Two points must be emphasized here: firstly, this model explicitly refers to 'documentation'—that is, the body of written texts produced in the course of the whole process of system development. Secondly, software development is defined as a process involving much more than the mere writing of computer programs; therefore, it extends over a longer time span. As for the first point, it is worth noting that the definition of 'documentation' will remain an open problem of software engineering for decades, and that the term will continue to shift from 'user documentation'—that is, all the manuals and guidelines provided to final users as an explanation of the software system—to 'technical' or 'internal documentation'—which functions both as a description of the system and as a means of communication between software developers. Regarding the second point, Selig comments that, although programming has traditionally been viewed as the production of code, in practice programmers perform their activities over the time span that goes from the moment when the 'problem' has been understood to the moment when the system becomes obsolete.

Both of these points concur in elucidating how the formalization of the process of software development is achieved in software en-

gineering. In order to control it, the process of software development is broken up into periods of activity, each of which produces,
as a result, a piece of writing. Each piece of writing is the point of
departure for the following phase, and it is supposed to be used as a
tool for developing more pieces of writing. Thus, the organization
of time is carried out through a practice that I call 'writing' because
this is the term used throughout the Garmisch conference report
and because it produces written texts—although the nature of such
'writing' needs to be explored further. Some of these written texts
are called 'documents', some are called 'software' or 'programs' or
'code', and these terms keep shifting. Moreover, the written text
resulting from the stage of analysis—namely, the description of the
problem—constitutes the point of departure for the whole process
of software development. At the same time, by describing the problem as something preexisting 'out there', this document also constitutes a narrative of the origin of the software system itself. More
precisely, it projects a 'problem' in the world to allow the software
system to become its 'solution'—or, it can even be said, it performs
the expulsion of the software system *as* a problem in order to justify
its very existence as a solution. By doing this, software engineering
also constitutes software as 'instrumental'—that is, as the means by
which the goal of solving a preexistent problem can be reached. At
the same time, though, software escapes instrumentality because
the distinction between problem and solution does not hold. As I
have just pointed out, software is both problem and solution—indeed, it emerges at the point where the distinction between the two
becomes undone and it exists only as the precarious stabilization of
this distinction.

Further on in the Garmisch conference report, Willem L. Van der
Poel formulates the following question:

> The specifications of a problem, when they are formulated pre-
> cisely enough, are in fact equivalent to the solution of the prob-
> lem. So when a problem is specified in all detail the formulation
> can be mapped into the solution, but most problems are incom-
> pletely specified. Where do you get the additional information to
> arrive at the solution which includes more than there was in the
> first specification of the problem?[47]

This passage shows the difficulty of describing the transition
from problem to solution purely in logical terms—those of the logical 'completeness' (or exhaustiveness) of the description of the
problem. Interestingly, Dijkstra answers Van der Poel's question by

comparing his own experience as a computer sciences teacher to that of a teacher of composition at a music school. One cannot teach creativity, he claims, nor can one ensure that one gets thirty gifted composers out of thirty pupils. What a teacher can do is to make pupils 'sensitive to the pleasant aspects of harmony'—but 'the rest they have to do themselves.'[48] Thus, Dijkstra resorts to individual creativity—or, as Derrida would have it, 'genius'—as an explanation for what is in excess of a procedural method and constitutes a leap beyond the programmable.[49] Recourse to subjectivity is Dijkstra's answer to the conceptual impasse arisen by the impossible transition from problem to solution. In fact, such unthinkable passage masks the expulsion of the problem from the process of software development in order to establish a narrative of the origins of the software system. Even more importantly, by separating problem from solution, while subsequently relating them through a series of written texts, the Garmisch conference report invests 'writing' with a central role in the organization of the time of software development.

The different written texts produced in different stages of software development take different and shifting forms and names. A complex relationship exists between the first of these texts—namely, the so-called external specifications (or 'specifications' *tout-court*) of the software system—and the text produced subsequently—that is, the 'internal design' (or simply 'design') of the system. Selig writes:

> External specifications at any level describe the software product in terms of the items controlled by and available to the user. The internal design describes the software product in terms of the program structures which realize the external specifications. It has to be understood that feedback between the design of the external and internal specifications is an essential part of a realistic and effective implementation process. Furthermore, this interaction must begin at the earliest stage of establishing the objectives, and continue until completion of the product.[50]

What has been characterized earlier on in the report as the 'how' and the 'what' of the system—or, in Selig's terms, 'analysis' and 'design'—is renamed here as the 'external specifications' and 'internal design', at the same time establishing a feedback loop between the inside and the outside of the software system. Selig's formulation implies a spatial representation of the software system, whose inner and outer parts are meant to be 'described' by written texts.

Furthermore, the external part of the software system is defined in terms of 'availability' (to the user) and 'control' (by the user)—that is, in terms of the description of software as a tool. Conversely, the internal design is defined as a text that provides a description of the programs—that is, of the single pieces of software—that will compose the software system as a whole. Selig uses here the term 'to realize': for him the specifications describe the system, while design and code 'realize' it. Moreover, a certain level of interaction ('feedback') between the specification and the 'realization' of the system is unavoidable, and actually essential for the system to progress towards completion.

Numerous remarks are made during the Garmisch conference report on the way in which the 'realization' of the software system must take place. In his paper entitled 'On the Interaction between Software Design Techniques and Software Management Problems', Kolence observes that the realization of the software system involves the 'interaction' between design techniques and management skills, which, however, cannot be easily separated from each other.[51] And yet, some kind of separation must be established if software engineering is to have a methodology—that is, a stable basis for the monitoring of the process of software development. As Alexander d'Agapeyeff notices, a monitoring methodology should substantially 'make software visible' in terms of programs, of 'the flow of their execution', of the 'shaping of modules', of their testing environment, and finally of the 'simulation of run time conditions'.[52] The only part of the software system and of the process of its development—or, even better, of the software system *in* development—which is actually visible is the part that is put in writing. Thus, during its development, the system is made visible through the written texts that mark the successive stages of such development. This process of 'making visible' and/as the monitoring of the software system is necessary if software development wants to move on from 'artistic endeavour' to scientific management. For d'Agapeyeff, the term 'artistic' has pejorative connotations—it functions as a synonym for 'unmonitorable', 'unmanageable', 'not reducible to a methodology', and even hints at an excessive individuality of the activity of programming. Kinslow comments:

> There are two classes of system designers. The first, if given five problems will solve them one at a time. The second will come back and announce that these aren't the real problems, and will eventually propose a solution to the single problem which

underlies the original five. This is the 'system type' who is great
during the initial stages of a design project. However, you had
better get rid of him after the first six months if you want to get a
working system. [53]

So, different mentalities are required by different stages of software development and an idiosyncratic approach to software, however brilliant, can put the whole process at risk. This brings us back
to the moment when Dijkstra takes recourse to individual creativity
in order to explain how software development progresses. Dijkstra's and Kinslow's positions interlock in a rather problematic way.
On the one hand, individual creativity is evoked as the mysterious
agent that allows for the translation of problem into solution, and
such translation is depicted as a form of the unexpected—analogous
to the unforeseeable and inexplicable leap that turns a student of
music into a talented composer. On the other hand, the view of
technology proposed by Kinslow, and quite likely shared by his
colleagues at the Garmisch conference, attributes the emergence of
the unexpected to the human mind: too much genius (or too many
unexpected consequences) can be dangerous, therefore it must be
kept under control, and a good project manager assigns the appropriate practitioners to the appropriate stage of the software project
in order to keep it manageable. Thus, in the process of software
development, the emergence of the unexpected seems to be both the
propeller of development itself *and* what puts development at risk.
Moreover, the transition from problem to solution is represented as
a 'good' form of the unexpected—that is, as an unforeseeable creative leap that, although it cannot be anticipated, is nonetheless manageable. The 'excessive' creativity of the 'system type' is in turn
portrayed as 'bad' unexpected—something that exceeds the management of the project and threatens it. Therefore, once again we are
faced with the 'point of opacity' of the unexpected consequences of
technology and with the impossible expulsion that software engineering tries to perform: while the aim of software engineering
claims to be the expulsion of the unexpected from technology, the
unexpected—represented as a creative leap—is also acknowledged
as constitutive of technology (and temporarily ascribed to human
'creativity'). Therefore, the question about the different kinds of
texts produced during the process of software development can be
reformulated as such: what is the relationship between these different texts and the unexpected? Could it be that such texts are differ-

ent precisely because they entertain different relations with what are represented here as the unforeseeable effects of technology?[54]

To understand this point better, it is worth noticing how, in the Garmisch report, J. A. Harr attempts to clarify the way in which the technical characteristics of the software systems and the organizational aspects of the working group mirror one another. Harr breaks down the design and production process into thirteen steps, starting from the extensive specification of all the system functions up to the final global testing of the system after it has been completed. He then proposes a structure of the programming group that reflects the structure of the system by allocating specific groups of programmers to each of its parts. One must be reminded here that, although the term 'structured programming' had yet to be invented, in 1968 so-called 'modular programming' was already in vogue—that is, an approach that allowed software developers to break down a complex system into different subsystems, each of which was supposed to perform a certain function and whose combination achieved the general purpose of the system.[55] Each subsystem could in turn be broken down into smaller components, until reaching a level where the whole system was decomposed into manageable parts—each a relatively independent piece of software that could be developed by single programmers separately. Randell calls this the 'divide and conquer approach to system complexity'.[56] The smallest components of the system, managed by individual programmers, are called 'modules', which is a synonym for 'computer programmes'— that is, aggregates of statements written in a programming language.[57] On the other hand, it was widely acknowledged that the structure of the software system reflected the structure of the programming group. For instance, Conway had already published his paper titled 'How Do Committees Invent?', where he formulated what later became known as Conway's Law—that is, that 'organizations which design systems are constrained to produce designs which are copies of the communication structures of the organizations.'[58] According to Conway,

> [a] contract research organisation had eight people who were to produce a COBOL and an ALGOL compiler. After some initial estimates of difficulty and time, five people were assigned to the COBOL Job and three to the ALGOL Job. The resulting COBOL compiler ran in five phases, the ALGOL compiler ran in three.
> Two military services were directed by their Commander-in-Chief to develop a common weapon system to meet their respec-

tive needs. After great effort they produced a copy of their organ-
isation chart. [59]

In other words, the structure of the software system and that of
the programming group reflected one other, with neither of them
assuming a determining role. Rather, it could be said that such
structures co-emerged during the process of software development.

This being the context, it comes as no surprise that the Garmisch
report contains an extensive discussion of the difficulty of breaking
down the process of software development and of naming its different stages, and that the overall terminology continues to shift between 'documentation', 'software', 'module', 'program', 'code', and
'writing'. The differences between the texts produced by programmers seem to be related to their belonging to different stages of
software development—that is, to the organization of time within a
specific project, which in turn is reflected by the structure of the
project group. To understand these differences better, let me now
turn to a later stage of software development, named 'implementation' (or 'production'), in which the texts produced in the phase of
design (which are generally written in some kind of formalism combined with natural language) are transformed into 'code'. Naur and
Randell clarify how implementation does not indicate 'just the making of more copies of the same software package (replication), but
the initial production of coded and checked programs'. [60] Naur
writes:

> Software production takes us from the result of the design to the
> program to be executed in the computer. The distinction between
> design and production is essentially a practical one, imposed by
> the need for a division of the labor. In fact, there is no essential
> difference between design and production, since even the pro-
> duction will include decisions which will influence the perfor-
> mance of the software system, and thus properly belong in the
> design phase. For the distinction to be useful, the design work is
> charged with the specific responsibility that it is pursued to a
> level of detail where the decisions remaining to be made during
> production are known to be insignificant as to the performance
> of the system. [61]

In this extremely important passage, Naur denies any essential
difference between design and implementation (production). For
him, the decision to break down the process of software development in different stages marked by different pieces of writing has
been made simply in order to enforce the division of labour. [62] Naur

in fact introduces a very subtle point here: the level of detail of the system description produced at the stage of design has the predominant goal of maintaining such a division of labour. More precisely, he seems to imply that an appropriate level of detail in design would completely erode the space left for implementers to make decisions of their own, and would thus destroy their ability to influence the software system in any way. The detail given in the piece of writing produced during the stage of design ultimately constitutes the means for the control of the workforce. It is even clearer now why for d'Agapeyeff a less and less systematic view of the software system—a kind of blindness, almost, or at least the selective forgetting of parts of the system—seems to be preferable at the later stages of the project in order to limit the risks of excessive 'creativity'.

But the most important point is Naur's acknowledgement that there is actually no essential difference between design and production. Thus, it can be argued that there is no difference between the written texts produced in these two stages—that is, between design documents (written in natural language and accompanied by formal notation) and computer programs. In fact, Naur rejects the existence of any intrinsic difference between what Hayles calls 'writing' and 'code'.[63] The only difference that Naur acknowledges is the introduction of a foreclosure—namely, at the level of design the 'practical' division of labour forecloses the possibility of decision at later stages of software development. What is already inscribed, or written down, at the stage of design cannot be changed at the stage of production; it cannot be decided upon any more; it cannot be undone, unmade; it is subtracted from the process of decision, of change, of further inscription. Writing performs a foreclosure of time through what is already inscribed—that is, the 'level of detail' of the design documentation. By foreclosing the possibility of a decision, writing also forecloses responsibility in the stage of production. Finally, it prevents feedback from production into design, from the 'how' into the 'what'—a feedback that is, nevertheless, necessary and even unavoidable, as we have seen earlier on when examining Selig's model of the software life cycle. Thus, the differentiation of the practice of writing in software development is an attempt at foreclosing an unavoidable iteration. For this reason, such a foreclosure is destined to fail—and yet, the process of software development relies on it and tries to preserve it at all times. The 'opacity' of the distinction between 'writing' and 'code' becomes apparent here: if it did not progress from specifications to code, the system would never be realized. And yet, there is no 'essential' difference between the two.

In fact, software development *must* involve a certain degree of iteration. Kinslow's contribution on this point is worth quoting at length here:

> The design process is an iterative one. . . . In my terms design consists of:
>
> 1. Flowchart until you think you understand the problem.
> 2. Write code until you realize you don't.
> 3. Go back and re-do the flowchart.
> 4. Write some more code and iterate to what you feel is the correct solution.
>
> If you are in a large production project, trying to build a big system, you have a deadline to write the specifications and for someone else to write the code. Unless you have been through this before you unconsciously skip over some specifications, saying to yourself: I will fill that in later. You know you are going to iterate, so you don't do a complete job the first time. Unfortunately, what happens is that 200 people start writing code. Now you start through the second iteration, with a better understanding of the problem, and it is too late. This is why there is version 1, version 2, version N. If you are building a big system and you are writing specifications you don't have the chance to iterate, the iteration is cut short by an arbitrary deadline. This is the fact that must be changed. [64]

Briefly put, for Kinslow, software professionals understand the system *through repeated attempts to build the system itself*. Drawing a flowchart is a first attempt to understand the system. On the basis of this flowchart, some code is developed which constitutes a second, further attempt to understand the system. At a certain point in time, the person writing the code (who may or may not be the one who drew the flowchart) meets some unexpected obstacle—typically, the person realizes that the flowchart contains some inconsistency. Therefore, the second understanding of the system feeds back into the first, causing it to be repeated differently and to produce a third understanding of the system (again in the form of a flowchart)—and so on. Not only is the process of understanding presented as a matter of 'making visible' here (for instance, code makes visible some inconsistencies or errors in the flowchart), but also, if the person going through the iteration cycles is one and the same, then that person's understanding of the system—what in philosophical terms we might call his or her 'consciousness' of the system—develops through the inscription of marks (flowcharts, code). In other words, the very process of the exteriorization of the system through writing makes the system understandable. I would go so far as to say that the different kinds of inscription illustrated in Kinslow's passage are the way in which the consciousness of the programmer is exteriorized *in/as* software.

A corollary of Kinslow's argument is that iteration puts the unforeseen consequences of technology at work in order to control technology: for instance, an inconsistency in the flowchart becomes unexpectedly apparent in code, in turn allowing for the modification of the flowchart itself. In fact, this example shows how unpredictability works at a very subtle level in software production—that is, it plays a pivotal role in enabling the passage between its different stages. However, although in this passage Kinslow argues in favour of iterability, he also acknowledges that in large projects which involve a huge number of participants it is quite difficult to iterate the process of design. Drawing on his own experience, Kinslow emphasizes that one tends to 'skip' parts of the specification because one 'unconsciously' postpones the understanding of such parts to the future. However, the subsequent development of code makes visible this gap in understanding—that is, it makes visible the unanticipated consequences of the software system which is being developed. Hence the 'danger', or risk, implicit in breaking down the process of software development into separate stages— danger which is nevertheless unavoidable if software development is to progress from specifications to code.

In sum, the different forms of material inscription produced in the process of software development—from specifications written in natural language to code—differ only in terms of their foreclosure of time, but such a foreclosure is ultimately impossible. The unexpected consequences of software cannot be avoided in the last instance. Indeed, they need to exist in order for software to exist, but they also must be excluded—continually and strenuously—in order for software to reach points of stability (for instance, its existence in the form of specifications, or flowcharts, or code).[65] On this point, Douglas T. Ross comments further:

> The most deadly thing in software is the concept, which almost universally seems to be followed, that you are going to specify what you are going to do, and then do it. And that is where most of our troubles come from. The projects that are called successful,

have met their specifications. But those specifications were based
upon the designers' ignorance before they started their job. [66]

The central problem of software development is thus the impossibility of following a sure path in which a system is completely and
exhaustively specified *before* it can be realized. The software professionals taking part in the Garmisch conference had a very clear view
of this 'paradoxical' aspect of software engineering. The specification of a system is never complete and inconsistencies become clear
only later on, when the process of implementation starts. Inconsistencies made visible by implementation in turn require an iterative
improvement, or rewriting, of specifications. In fact, what Ross is
saying here is that the sequencing of the process of software development over time, which is the basis of software engineering, is
impossible *as such*. One invariably starts doing what one wants to
do before knowing what it is. A project is successful because it
meets its specifications — that is, it does what its specifications say it
must do — but the specifications were written when one did *not* actually know what the system was supposed to do. How can something based on the lack of knowledge be realized successfully? This
paradox clarifies the particular understanding of time that programmers develop in their interaction with software. Not only does
one find out what the system does only by constructing it; the original ignorance of what the system does is *constitutive of* the system.
One invariably starts neither completely knowing nor being completely ignorant about what the system will do — and both knowledge and ignorance are made visible through the very process of
starting anyway. The act of writing specifications for the system
gives shape to the system by making some of its unforeseeable consequences visible. Infelicity is constitutive of the possibility of felicity. [67] In a way, this is how software constitutes the time of its own
development: because, where does one start *from*? One always starts
albeit one does not know what will be. The system one designs now
will become the past of the system one will have realized. *Such a
system is unforeseeable — but one starts anyway, and this constitutes both
the present and the future of the system.*

However, one must be wary of the idea — suggested by Kinsley
and other conference participants — that the closer one is to the stage
of 'implementation', that is, of code, the more 'exteriorized', and the
less unpredictable, the consequences of software become. I want to
argue instead that the exteriorization of the software system always
brings forth new and different possibilities of unforeseen conse-

quences. The concept of 'notation', as it is debated in the Garmisch conference report, becomes quite significant here. Early in the report the importance of having a formal notation for design is pointed out. Such notation would be the equivalent of Boolean algebra for computer processor design. In our own terms, notation can be viewed as one of the possible modes of inscription of software — of which specifications written in natural language, graphical flowcharts, and computer programs are all instances. To give but one example, the following sequences of letters and numbers

> Copy the value contained in the memory location 202 into the register AX
> MOV AX,[202]
> A10202
> 10100001000000001000000010

are four equivalent notations for the same state of polarities on a magnetic disk. They are expressed in natural language, Assembler language, hexadecimal machine language, and binary machine language, respectively. They are all part of software, since they are all notations used in software development.

However, as Kolence argues, the best notation for system design is the one that 'permits an initial description of the internal design of software to be broken apart into successively more detailed levels of design, ultimately ending up with the level of code to be used'.[68] Here Kolence attributes to notation a significant role in the sequencing of time. In fact, he describes the whole process of software development, and the software system itself, as a process of inscription, and thus of linearization, in which different kinds of notation allow for different ways of sequencing time. Furthermore, he argues that '[s]oftware design notation . . . should decompose naturally from the highest level of design description down to design documents which suffice for maintenance of the final software.'[69] Thus, the sequencing of notations in the process of software development coincides with steps in time—that is, it is a sequenced process of exteriorization. Code can well be the most advanced stage of software development, but it does not constitute the complete foreclosure of the unexpected consequences of software.

But in what way does the unexpected keep emerging in code? To understand this point, it is important to examine the so-called principle of extensibility. In the Garmisch conference report, extensibility is defined as the possibility of extending the functionalities of the system by reprogramming parts of it (for instance, by adding pieces of software) after the system has been completed and delivered to its users. The capacity of the system to allow for its own extension in the future—or for the broadening of its own life cycle—is based on the characteristics of the notation in which the system is exteriorized. It can be said that the system must extend into the future as far as it can and as fast as it can, thus capturing and calculating time, while at the same time leaving a gate to the unforeseeable open. Extensibility is related to the modularity of software—that is, the possibility to isolate functional elements of the system that can be taken out, reprogrammed, and put back in. Importantly, the Garmisch conference report relates extensibility to the concept of modularity. In his conference paper entitled 'Aids in the Production of Maintenable Software', H. R. Gillette adds that accurate 'documentation' should be maintained—that is, an accurate description of what the system does and of how it does it—in order for future developers to be able to understand it and modify it.[70] Once again, the unexpected consequences of software are inscribed in software itself in all its forms; in particular, they are inscribed in code by means of extensibility and modularity. The combination of extensibility and modularity constitutes a way to calculate the future— but at the same time, since nobody can anticipate what an open-ended system will do, or what can be done with it, it also keeps the possibility of the unforeseeable open.

Gillette's use of the term 'documentation', which refers to those written texts that help the final user understand the software system—generally called 'user manuals'—is extremely relevant, because it is here that the figure of the 'user' makes its appearance in the Garmisch conference report. I want to argue that in the Garmisch report the 'user' is actually a name given to a part of the process of software design. In the figure of the user, both the instability of the instrumental understanding of software and the capacity of software to escape instrumentality through the unexpected consequences it generates become apparent.

The user emerges in the report as a problematic figure towards which software developers have ambivalent feelings. On the one hand, J. N. P. Hume suggests that designers must not 'over-react' to individual users' requests—that is, in order to develop an effective and usable software system, they must identify the requirements 'common to a majority of users' and focus on them.[71] On the other hand, J. D. Babcock argues for the intelligence of the users. He comments: '[t]he users are the people who do our design, once we get started.'[72] In doing so, Babcock attributes to 'the users' an essential role in the process of software development various decades before the emergence of so-called user-centred design models of the late 1980s.[73] In a sense, the 'user' is inscribed in these theories and practices not just as an 'idealization' or a 'function' of the system, but as a field of forces constitutive of the whole process of software development.[74] However, the conference participants express a general discomfort in interacting with 'the user'. Manfred Paul depicts the user as someone who 'does not know what he needs', and he couples this with another kind of ignorance: users are actually 'cut off from knowing what is or what might be available'.[75] And Al Perlis adds: 'Almost all users require much less from an operating system than is provided.'[76] In these two passages users are understood alternately as unable to understand their own needs—and thus unable to pose clear requests to technology—and overwhelmed by the technological offer—and thus incapable of making the most of the functionalities provided by technology. These complaints about 'users' are a familiar feature not just of software engineering, but of the general approach of software developers to their non-technical counterparts.[77] However, it would be reductive to interpret such complaints merely in terms of the difficulties encountered by software practitioners in communicating with non-technical users. Importantly, J. W. Smith notices that designers usually refer to users as 'they', 'them'—a strange breed living 'there in the outer world, knowing nothing, to whom nothing is owed'. He also adds disapprovingly that most designers 'are designing . . . for their own benefit—they are literally playing games'—they have no conception of validating their design—or at least of evaluating it in the light of potential use.[78] This representation of the user as someone 'out there'—someone whose 'needs' should be taken into account in order to validate software instrumentally—is particularly relevant if we want to understand how the figure of the user operates in software engineering. In fact, I want to suggest that the 'user' and their 'needs' are part of the fictional 'origin' of the software system. As I have shown earlier on, in the Garmisch conference report, 'society' is the locus of a projection of the 'demands' that are supposedly made to technology—while at the same time a preexisting 'problem' is projected in the world 'out there' in order to justify the existence of software. Here I want to emphasize how the figure of the user plays an analogous role—that is, the user's needs are part of a narrative that software developers construct in order to justify the system they are developing. This is not to say that users do not really exist or that they do not express their demands in terms of what functionalities should be provided by a software system. In fact, the Garmisch conference report takes the communication with users very seriously at all levels. And yet, the figure of the 'user' is positioned by the report outside the process of software development in a constant and incomplete movement of 'expulsion' of certain characteristics of software *as* 'user needs'. In Goos's words, software developers need to 'filter the recommendations coming from the outside'.[79] A double strategy is at work here, which acknowledges the importance of users while focusing on how to keep them at bay. Randell even laments the amount of time wasted on 'fending off the users'.[80] Thus, 'the user' is both constituted and neutralized through various practices of writing: while it is acknowledged that software development is set in motion by the very existence of (potential) users and that it needs their feedback, the very inscription of the software system in terms of specifications, design, code, and user manuals acts as a form of containment of the (supposed) user's exigencies.

Even more importantly, as I have suggested above, the figure of the user is associated with the extensibility of software. Not only should a software system be 'open-ended', but also 'documentation' must be provided to users—that is, written texts whose goal is 'to train, understand, and provide maintenance'.[81] User manuals are what enables users to enter an active relationship with software; they allow users to engage with a system whose open-endedness is inscribed in code. Therefore, documentation also constitutes a point where the capacity to take advantage of such open-endedness and to take the system into an unexpected direction is ultimately handed over to the users. This does not mean that any user can actively reprogram any system. In fact, according to the Garmisch conference report, one of the aims of software developers is to make the system 'dumb-proof'—that is, robust and resilient enough to resist 'improper' uses on the part of inexperienced and non-technical users.[82] And yet, it seems to me that the figure of the user is the locus where the instrumentality of software is both reasserted by implicitly defining it as a tool to be 'used' and opened up to unexpected consequences. As I will show in Chapter 4, the ambivalent figure of the user will be at the core of many unexpected developments of software engineering in the 1980s and 1990s.

NOTES

1. Michael S. Mahoney, "Finding a History for Software Engineering," *Annals of the History of Computing* 26, no. 1 (2004): 8.

2. Watts S. Humphrey, "The Software Engineering Process: Definition and Scope," in *Representing and Enacting the Software Process: Proceedings of the 4th International Software Process Workshop, 1989* (ACM Press, 1989b), 82.

3. Mary Shaw, "Prospects for an Engineering Discipline of Software," *IEEE Software* 7, no. 6 (1990): 15.

4. Mahoney, "Finding a History for Software Engineering," 8.

5. Mahoney, "Finding a History for Software Engineering," 8.

6. Peter Naur and Brian Randell, eds., *Software Engineering: Report on a Conference Sponsored by the NATO Science Committee, Garmisch, Germany, 7th to 11th October 1968* (Brussels: NATO Scientific Affairs Division, 1969), 13.

7. Mahoney, "Finding a History for Software Engineering," 8.

8. Naur and Randell, eds., *Software Engineering: Report.*

9. The report of the first NATO Conference on Software Engineering, held in Garmisch from 7th to 11th October 1968, was edited by Peter Naur and Brian Randell soon after the conference (Naur and Randell, eds., *Software Engineering: Report*). NATO was in charge of the actual printing and distribution, and the report became available three months after the conference, in January 1969. The report of the second conference, held in Rome from 27th to 31st October 1969, was edited by John Buxton and Brian Randell and published in April 1970. (John N. Buxton and Brian Randell, eds., *Software Engineering Techniques: Report on a Conference Sponsored by the NATO Science Committee, Rome, Italy, 27th to 31st October 1969* [Birmingham: NATO Science Committee, 1970]). Both reports were later republished in book form. (John N. Buxton, Peter Naur, and Brian Randell, eds., *Software Engineering: Concepts and Techniques* [New York: Petrocelli-Charter, 1976]). In 2001, Robert M. McClure made both reports available for download in .pdf format at http://homepages.cs.ncl.ac.uk/brian.randell/NATO/ . The pagination of the .pdf version slightly differs from the original printed version. All the references made in this chapter are based on the original pagination.

10. Cf. Gary Hall, "IT, Again: How to Build an Ethical Virtual Institution," in *Experimenting: Essays with Samuel Weber*, ed. Simon Morgan Wortham and Gary Hall (New York: Fordham University Press, 2007), 116–40.

11. Brian Randell, "Software Engineering in 1968," *Proceedings of the IEEE 4th International Conference on Software Engineering* (Munich, Germany, 1979), 1. The Garmisch conference report describes the participants as 'about 50 experts from all areas concerned with software problems—computer manufacturers, universities, software houses, computer users, etc.' (Naur and Randell, eds., *Software Engineering: Report*, 13). The fact that the participants had quite heterogeneous backgrounds was in line with the intentions of the organizing committee (Randell, "Software Engineering in 1968," 1). Nearly half of them came from North America, the rest from various European countries, and among them were outstanding scholars and practitioners, many of whom were meeting each other for the first time. Many went on to write fundamental works in software engineering in the following decades. A list of participants is given in Naur and Randell, eds., *Software Engineering: Report*, 212–17.

12. Randell, "Software Engineering in 1968," 1.

13. Naur and Randell, eds., *Software Engineering: Report*, 121.

14. Naur and Randell, eds., *Software Engineering: Report*, 3.

15. Randell, "Software Engineering in 1968," 5. In the late 1960s, the unit of measure for determining whether a system was 'large' was the number of lines of code it contained. A large software system could include several thousand lines of code. Another popular unit, still used nowadays, was the 'man-year', that is the number of years an average programmer would spend on the system if he or she were to develop that system by themselves. A few years after the Garmisch conference, a subunit of the man-year—the man-month—became the title of a classic of software engineering, Frederick Brooks, *The Mythical Man-Month: Essays on Software Engineering, Anniversary Edition* (Reading, MA: Addison-Wesley, 1995).

16. Joseph C. R. Licklider, "Underestimates and Overexpectations," in *ABM: An Evaluation of the Decision to Deploy an Anti-Ballistic Missile System,* ed. Abram Chayes and Jerome B. Wiesner (New York: Signet, 1969), 118.

17. Randell, "Software Engineering in 1968," 5.

18. A famous example of those earlier operating systems was OS/360, which ran on IBM computers of the 360 series. A vastly innovative operating system, OS/360 was extremely complex for the time, and, albeit becoming rapidly popular, it contained a number of technical flaws. Its development during the first half of the 1960s was characterized by many setbacks, mainly due to the poor management of time, and it was largely discussed during the Garmisch conference. Frederick Brooks became the project manager of OS/360 at IBM in 1964. In order to speed up the process of software development, he mistakenly added more programmers to the project, falling behind schedule as a result. Later on, drawing on this experience, he formulated the principle that 'adding more manpower to a late software project makes it later', which became known as 'The Brooks's Law' (cf. Brooks, *The Mythical Man-Month*). I will return to Brooks's contribution to software engineering in Chapter 4.

19. Laszlo A. Belady and Meir M. Lehman, "A Model of Large Program Development," *IBM Systems Journal* 15, no. 3 (1976): 230. A release is a 'stable' state of a software system—namely, a version of the system that is deemed complete and correct enough to be delivered to its users. Such delivery does not prevent system developers from improving and extending the system until they reach a new (and supposedly better) stable version of it. Releases are usually indicated by consecutive numbers. An instruction is usually a single line of code containing a basic command or data. The sequencing of instructions constitutes a program. Release 16 was a considerably large system for the time. In fact, at the 1968 conference E. E. David pointed out that OS/360 had by then absorbed five thousand man-years of work (Naur and Randell, eds., *Software Engineering: Report*, 15).

20. As we have seen in Chapter 2, a 'real-time' system is a software system which is able to respond to changes in its environment 'as soon as they happen'—that is, in a fast and effective way. Real-time systems tend to be complex, high-risk, low fault-tolerance systems.

21. Naur and Randell, eds., *Software Engineering: Report*, 15.

22. Naur and Randell, eds., *Software Engineering: Report*, 15.

23. Bernard Stiegler, *Technics and Time, 1: The Fault of Epimetheus* (Stanford, CA: Stanford University Press, 1998), 17.

24. Naur and Randell, eds., *Software Engineering: Report*, 15–16.

25. Jacques Derrida and Bernard Stiegler, *Echographies of Television: Filmed Interviews* (Cambridge: Polity Press, 2002), 45.

26. Naur and Randell, eds., *Software Engineering: Report*, 15.

27. Naur and Randell, eds., *Software Engineering: Report*, 16.

28. Naur and Randell, eds., *Software Engineering: Report*, 17.

29. Naur and Randell, eds., *Software Engineering: Report*, 16.

30. Mary Shaw, "Remembrances of a Graduate Student," *Annals of the History of Computing, Anecdotes Department* 11, no. 2 (1989): 141.

31. Naur and Randell, eds., *Software Engineering: Report*, 18.

32. Cf. Jürgen Habermas, *The Theory of Communicative Action* (Cambridge: Polity Press, 1991).

33. Cf. Jacques Derrida, "Letter to a Japanese Friend," in *Derrida and Différance*, ed. Robert Bernasconi and David Wood (Warwick: Parousia Press, 1985).

34. Cf. Bertrand Gille, *History of Techniques* (New York: Gordon, 1986); Stiegler, *Technics and Time 1*.

35. Naur and Randell, eds., *Software Engineering: Report*, 17.

36. Naur and Randell, eds., *Software Engineering: Report*, 17.

37. Randell, "Software Engineering in 1968," 2. See also Wendy Hui Kyong Chun, *Programmed Visions: Software and Memory* (Cambridge, MA and London: MIT Press, 2011).

38. Randell, "Software Engineering in 1968," 2. See also Paul Ceruzzi, *A History of Modern Computing* (Cambridge, MA and London: MIT Press, 2003).

39. Naur and Randell, eds., *Software Engineering: Report*, 150.

40. Naur and Randell, eds., *Software Engineering: Report*, 17.

41. Naur and Randell, eds., *Software Engineering: Report*, 17.

42. Cf. Jacques Derrida, *Of Grammatology* (Baltimore: The Johns Hopkins University Press, 1976); André Leroi-Gourhan, *Gesture and Speech* (Cambridge, MA: MIT Press, 1993).

43. Mary Shaw, "Remembrances of a Graduate Student," 99.

44. Mary Shaw, "Remembrances of a Graduate Student," 99–100.

45. Naur and Randell, eds., *Software Engineering: Report*, 21.

46. Naur and Randell, eds., *Software Engineering: Report*, 21.

47. Naur and Randell, eds., *Software Engineering: Report*, 52.

48. Naur and Randell, eds., *Software Engineering: Report*, 52.

49. Cf. Jacques Derrida, *Specters of Marx: The State of the Debt, the Work of Mourning, and the New International* (New York and London: Routledge, 1994) and Timothy Clark, "Deconstruction and Technology," in *Deconstructions. A User's Guide*, ed. Nicholas Royle (Basingstoke: Palgrave, 2000), 249.

50. Naur and Randell, eds., *Software Engineering: Report*, 22.

51. Naur and Randell, eds., *Software Engineering: Report*, 24.

52. Naur and Randell, eds., *Software Engineering: Report*, 24.

53. Naur and Randell, eds., *Software Engineering: Report*, 24.

54. It is important to clarify that, in my deconstructive reading of the Garmisch report, I am deliberately avoiding an ontologization of what I have named 'the unforeseen consequences of technology' or 'the unexpected', and the conference participants also name 'risk', 'failure', 'creativity', and so on. What I want to point out is that, in the seminal text of software engineering, the 'unexpected' is inscribed as a part of the conceptual functioning of software—a force that constitutes and simultaneously undoes the process of software development and that can never be 'explained away'. In Chapter 5, we shall see in detail how the unexpected is inscribed in the theory of programming languages.

55. Randell, "Software Engineering in 1968," 4. Dijkstra's well-known paper "Notes on Structured Programming" was written in August 1969. However, at the time of the Garmisch conference, his letter "Go To Statements Considered Harmful"—which famously inaugurated the age of structured programming—as well as the first responses to it had already been published (Edsger W. Dijkstra, "Go to Statements Considered Harmful," *Communications of the ACM* 11, no. 3 [1968]: 146–48).

56. Randell, "Software Engineering in 1968," 4.

57. Larry L. Constantine, "The Programming Profession, Programming Theory, and Programming Education," *Computers and Automation* 17, no. 2 (1968): 15.

58. Randell, "Software Engineering in 1968," 5.

59. Melvin. E. Conway, "How Do Committees Invent?" *Datamation* 14, no. 4 (1968): 30.

60. Naur and Randell, eds., *Software Engineering: Report*, 30. To clarify Naur and Randell's curious remark, it is worth mentioning that in the late 1960s the method for producing copies of software was not as consolidated and secure as it is today, neither was it completely automatized. In fact, it involved a lot of manual activities and it could frequently introduce errors in the piece of software being copied.

61. Naur and Randell, eds., *Software Engineering: Report*, 31.

62. The very distinction between hardware and software, which dates from John von Neumann and his understanding of the general structure of computers, is perhaps the most famous example of a technical decision coupled with a division of labour. Cf. Jay David Bolter, *Turing's Man: Western Culture in the Computer Age* (London: Duckworth, 1984).

63. Cf. Katherine N. Hayles, *My Mother Was a Computer: Digital Subjects and Literary Texts* (Chicago: University of Chicago Press, 2005), 16.

64. Naur and Randell, eds., *Software Engineering: Report*, 32.

65. It is worth noting that Kinslow's passage shows that the expression 'to write code' was very common in the late 1960s. For instance, Kinslow states that, if different people are involved in the process of iteration, different processes of writing start in different moments, drawing on different (and differently flawed) points of departure, which leads to the development of different versions of the software system. It is quite obvious from the Garmisch conference report that in 1968 the use of the term 'writing' in relation to software was largely accepted, although writing had still to be theorized as a way to make software visible and thus to control its development.

66. Naur and Randell, eds., *Software Engineering: Report*, 32.

67. Cf. Jacques Derrida, *Limited Inc.* (Evanston, IL: Northwestern University Press, 1988).

68. Naur and Randell, eds., *Software Engineering: Report*, 47.

69. Naur and Randell, eds., *Software Engineering: Report*, 50.

70. Naur and Randell, eds., *Software Engineering: Report*, 39.

71. Naur and Randell, eds., *Software Engineering: Report*, 40.

72. Naur and Randell, eds., *Software Engineering: Report*, 40.

73. On user-centred design, see Donald A. Norman, *The Design of Everyday Things* (New York: Basic Books, 1988).

74. For a critique of the 'idealization' of the user in computer interface, see Matthew Fuller, *Behind the Blip: Essays on the Culture of Software* (New York: Autonomedia, 2003). I have discussed Fuller's argument in Chapter 2.

75. Naur and Randell, eds., *Software Engineering: Report*, 40.

76. Naur and Randell, eds., *Software Engineering: Report*, 40.

77. Cf. Bolter, *The Turing Man*.

78. Naur and Randell, eds., *Software Engineering: Report*, 40.

79. Naur and Randell, eds., *Software Engineering: Report*, 41.

80. Naur and Randell, eds., *Software Engineering: Report*, 41.

81. Naur and Randell, eds., *Software Engineering: Report*, 39.

82. Naur and Randell, eds., *Software Engineering: Report*, 40.

FOUR

From the Cathedral to the Bazaar

Software as the Unexpected

> Any tool should be useful in the expected way, but a truly great tool lends itself to uses you never expected.
>
> —Eric S. Raymond

In the foundational texts of software engineering of the late 1960s, 'software' was constituted as a process of material inscription, an act of conceptualization that took place through the continuous undoing and redoing of the boundaries between 'software' itself, 'writing', and 'code'. In this chapter, I want to investigate how the mutual co-constitution of 'software', 'writing', and 'code' was established in the software engineering of the late 1970s and early 1980s. I want to argue that in this period software engineering reconfirmed its capacity for continually opening up and reasserting the instrumental conception of software—that is, the understanding of software as a tool. Software engineering kept questioning itself as a discipline and constantly reevaluated the validity of its methodologies—especially those deployed for time management. In this second phase of software engineering, even though the sequencing of the process of software development appeared consolidated, the distinction between 'software', 'writing', and 'code' remained unstable and 'software' continued to be perceived as intrinsincally fallible. Furthermore, in the mid-1970s, the problem of coordinating different 'minds'—that is, the individual software professionals

99

who took part in a single software project—made its appearance in software engineering.[1] If one views software in terms of the exteriorization of consciousness, it can be said that consciousness was made discrete, and divided into different, and sometimes conflicting, individual consciousnesses. The coordination between individual consciousnesses was understood in terms of 'communication'.

In order to investigate how the problem of software calculability was approached in the 1970s and 1980s, in this chapter I examine two of the fundamental texts on time management in software development, both written by Frederick Brooks in the mid-1970s and mid-1980s, respectively. The first one is Brooks's book entitled *The Mythical Man-Month*. Originally published in 1975, it rapidly became the most famous text on time estimates in the whole history of software engineering and remained extremely influential for at least two decades. However, ten years later, in 1986, Brooks published an article entitled 'No Silver Bullet', which also became a classic, where he revised his theses of 1975. Such a revision stimulated a heated debate among software engineering scholars and practitioners. An accurate analysis of Brooks's works of 1975 and 1986 is key to understanding in what way the calculability of time in software development was conceptualized during the late 1970s throughout the 1980s and early 1990s.

In the second section of this chapter, I contrast Brooks's understanding of the time of software development against the one proposed by the open source movement by looking at the foundational article of software engineering for open source programming—namely, Eric Steven Raymond's 'The Cathedral and the Bazaar'. Written in 1997, republished online many times, and later reworked in a book of the same title, this text constitutes Raymond's answer to Brooks's original argument on software development.[2] This analysis will lead me to view the emergence of open source programming in the 1990s, with its own brand of software engineering practices and theories, as one of the unforeseen consequences of the conception of software development of the 1970s and 1980s. The open source movement proposes a totally unexpected answer to the problem of coordination between various software designers. It does so by replacing software development as a process of communication with software development as a collective process of overlapping reinscriptions.

Fred Brooks became the project manager of the Operating System/360 (OS/360) at IBM in 1964. OS/360 contained many technical innovations and became largely popular, albeit it entailed—as was quite common at the time—a number of technical flaws.[3] The main problem encountered by the OS/360 project was related to time management. As Brooks would candidly recall ten year later, OS/360 'was late . . . , the costs were several times the estimate, and it did not perform very well until several releases after the first'.[4] Brooks left IBM in 1965, when he joined the University of North Carolina at Chapel Hill and began a careful investigation of what had actually gone wrong with the OS/360 development in 1964–1965. The result was *The Mythical Man-Month*, where he meticulously examined his experience and identified his fundamental mistake as a project manager in terms of wrong time estimates— also putting forward what later on would become known as 'The Brooks's Law' on time management.

The imaginative style deployed by Brooks in *The Mythical Man-Month* is replete with metaphors, many of which have a distinct biblical flavour—first and foremost the famous metaphor of software development as a 'cathedral', which in 1997 will be countered by Raymond with his metaphor of the 'bazaar.' However, Brooks opens his book with another well-known image—that of software as a 'tar pit'. He writes: 'large-system programming has over the past decade been such a tar pit, and many great and powerful beasts have thrashed violently in it. Most have emerged with running systems—few have met goals, schedules, and budgets.'[5] The 'tar pit' is nothing but a figure of time—or, more precisely, a figure of the failure of time management in software development. However, while the participants in the first NATO Conference on Software Engineering of 1968 view software technology as a constantly innovating field that advanced ever too fast and were desperate to find a way to manage the speed of software growth, quite strikingly Brooks's metaphor of the tar pit suggests that, if there is a problem with software development, it has to do with slowness rather than with excessive speed.[6] No single factor, he adds, seems to be uniquely responsible for the delays of software development: 'any particular paw can be pulled away' from the tar pit, but the accumulation and interaction of simultaneous factors 'brings slower and slower motion'. Brooks's image of time as a tar pit is also an image of sinking—or, more precisely, of slower motion: one moves slower and slower, until one sinks. Indeed, sinking seems almost unavoidable, since 'the fiercer the struggle, the more entangling the tar.'[7]

Brooks's aim in *The Mythical Man-Month* is to understand the tar pit in which OS/360 'sank' when he was its project manager and, more generally, to comprehend all the tar pits that seemed to threat-en the development of the majority of large software systems in the early 1970s. At the time of his writing, software development was already understood as an industrial process.[8] The industrialization of software, which was still under way in the late 1960s, is taken by Brooks as a *fait accompli*. He actually distinguishes 'programming' — the individual task of writing a program—from the development of a 'programming system product'—the industrial production of large software systems—and clearly establishes that he is interested in the latter. The very expression 'programming system product' emphasizes both the systematic and the industrial aspects of the kind of software that Brooks wants to examine. However, the activity of programming is part of the development of software products, and it is from the examination of this activity that Brooks starts. For him, no matter how fulfilling, creative, and rewarding, programming also presents many 'woes'. He writes: 'First, one must perform perfectly. The computer resembles the magic of legend in this respect. . . . If one character, one pause, of the incantation is not strictly in proper form, the magic doesn't work. . . . Adjusting to the requirement for perfection is, I think, the most difficult part of learning to program.'[9] Brooks identifies the perfection of the programmer's performance with the perfection of the piece of software he is writing. In fact, it is software that has to perform perfectly — namely, it has to conform to its own specifications which, as we already know, basically describe the required behaviour of the system. Software must perform as expected: it must be perfectly foreseeable. Brooks associates perfection with magic and identifies the good functioning of software with the good functioning of the spell—a spell in which everything is relevant: a misplaced space can determine a totally unexpected behaviour of the program when it is executed, thus failing to execute the spell perfectly. In other words, *a misplaced character makes time incalculable.*[10]

Even more important, and consistent with his metaphor of the tar pit, Brooks downplays the importance of the speed of software growth that just a few years earlier in Garmisch was considered such a big problem.[11] Brooks points out that, although software systems start becoming obsolete as soon as they are completed, newer products take time to be designed and implemented. Therefore, the obsolescence of an existing software product is not as fast as it might seem. Being time-consuming, the very activity of developing a 'real' software system imposes limits on the speed of technology. Since software was still undergoing a process of constant innovation in the mid-1970s, Brooks understands slowness first and foremost as the fact that projects tend to fall behind schedule in terms of what he refers to as 'calendar time'—that is, industrial time. The slowness of software is not a general slowness in the technological advancement of software; rather, it is software's resistance to conform to the deadlines of industrial production.

In sum, Brooks attributes software's 'slowness'—the 'tar pit'—to the ineffective management of time, and particularly to bad time estimates. And yet, he upholds the perfect management of time as the (impossible) ideal of software development. What Brooks is debating here is the fundamental problem of the industrial production of software—namely, the establishment of realistic time estimates, which in turns leads to the setting of reasonable schedules, so that the corresponding deadlines can be met. In large-scale software systems, he notices, a project manager and his staff are in charge of these decisions, and they make them on the basis of their experience. And yet, Brooks warns, 'all programmers are optimists.'[12] They always assume that there will be 'enough time'. In fact, this notion of 'optimism' is crucial for the understanding of Brooks's concept of software's slowness. 'Optimism' actually conveys the *impossibility of estimating the unexpected*. Programmers are incapable of calculating all the possible consequences of software development, including those that will require more labour than expected and that will therefore delay the completion of a project. Seven years after the Garmisch conference, Brooks's book shows that the question of time management is still unresolved, and although it is represented in quite a different way—that is, through the depiction of software as something that is advancing slower than expected—it is still related to the sequencing of the process of software development.

In order to understand how Brooks conceptualizes the sequencing of time development, it is worth noting that he views software development as a 'creative' process.[13] In *The Mythical Man-Month* he explicitly quotes Dorothy Sayers's book, *The Mind of the Maker*, which explores at length the analogy between human creation (especially literary creation) and the Christian doctrine of the Trinity.[14] Sayers divides any creative activity into the three stages of the 'idea', the 'implementation', and the 'interaction'—a division that will remain fundamental in Brooks's thought and that ten years later he will still support, by explicitly relating it to the Aristotelian separation between the essential and the accidental, the ideal and the material. It is worth noting here that what Brooks refers to as 'Aristotelian' is in fact the distinction between the ideal and the material established by Plato in the *Republic* (and then reflected in Aristotle's philosophy). Such a distinction is expressed in the well-known myth of the cave (in which Plato depicts ordinary human beings, deprived of philosophical education, as prisoners in a cave forced to look at shadows rather than at real things—by this implying that the ordinary world of sensible objects is far less real than the world of concepts or forms).[15] However, paraphrasing Sayers, Brooks writes:

> [a] book . . . or a computer, or a program comes into existence first as an ideal construct. . . . It is realized in time and space, by pen, ink, and paper, or by wire, silicon, and ferrite. The creation is complete when someone reads the book, uses the computer, or runs the program, thereby interacting with the mind of the maker.[16]

This passage clarifies that for Brooks the three stages of creation roughly coincide with the conception of a work of art or technology, its concretization in a material form, and its fruition on the part of viewers, readers and so on. But then Brooks adds, somewhat unexpectedly: 'for the human makers of things, the incompleteness and inconsistencies of our ideas become clear only during implementation. Thus it is that writing, experimentation, "working out" are essential disciplines for the theoretician.'[17] Notwithstanding his belief in the separation between the conception of a project and its realization, or between the conceptual and the material dimensions—which will become even clearer in his article of twenty years later, where he will introduce an even less tenable distinction between 'concept' and 'representation'—here Brooks seems to be aware that the process of software development is essentially one of 'exteriorization'. In turn, such exteriorization involves the iteration between different stages.

Brooks's distinction between conception and realization mirrors the one between software as a product and software as a process that is also constantly opened up and reaffirmed in the Garmisch conference report.[18] The system and its development are never clearly separable. For instance, the failures in the process of software development—failures that lead to time and investment spinning out of control—are attributed to the poor understanding of the software system in the first place. On the other hand, understanding the system is presented as a matter of 'making it visible': for instance, 'code', or the implementation of the software system, can make visible inconsistencies or errors in a flowchart—that is, in the specifications of the system. For this reason, some iteration between the stages of specification and implementation is necessary.

To a certain extent, as I have noticed above, Brooks acknowledges that the process of the 'exteriorization' of the system makes its inconsistencies visible. Indeed, he seems to concede that software exists *only* as exteriorization. The distinction between 'the ideal' and its 'realization' seems therefore to become undone at the very moment of its establishment. In fact, Brooks appears quite unsure as to how to keep the two aspects apart. Therefore, his argument encounters a major difficulty—or better, a point of opacity—here. On the one hand, he states that the 'realization' of the system takes time because of the difficulty of manipulating the 'physical media' involved, while on the other hand he attributes such a difficulty to inconsistencies that can be found in the very conception of the system. From this second point of view, difficulties would ultimately reside at the 'ideal' level, and the 'realization' of the system would actually coincide with its conceptual clarification.

Brooks introduces further ambiguity into his argument by surprisingly claiming that the medium of programming is 'an exceedingly tractable medium'. He explains: '[programmers] build from pure thought-stuff: concepts are very flexible representations thereof. Because the medium is tractable, we expect few difficulties in implementation; hence our pervasive optimism. Because our ideas are faulty, we have bugs; hence our optimism is unjustified.'[19] Thus, Brooks suggests that programmers' optimism—that is, their belief in the calculability of time—relies on their faith in the tractability of the whole system (both as a concept and as its realization) and that, in so doing, they overlook the system's inconsistencies. In sum, Brooks places the difficulty of developing a software system alternatively at the 'material' level, albeit this is contradicted by the supposed tractability of the computer as a support for inscription, and at the 'ideal' level, therefore making 'realization' (or 'implementation') the place where difficulties are actually clarified. This ambivalence suggests that the separation between conception and realization is ultimately untenable.

To recapitulate the above argument, although the distinction between the conception of a software system and its material realization (or implementation) appears quite unstable, Brooks invokes precisely the Platonic distinction between the ideal and the material in order to provide a foundation for his model of the sequencing of time in a software project. At this point, the crucial problem of Brooks's theory of time management emerges—namely, the 'mythi-

cal man-month'. Man-time units of measure—the man-month, the
man-year, and even the man-millennium—were already in use in
the 1960s to measure the size of software systems and the duration
of software projects. These units are defined respectively as the
number of months or years that an average programmer would
spend on the system if they were to develop that system by themselves.[20] However, Brooks argues that the man-month is a fallacious
unit of measure that leads to erroneous time estimates. The cost of a
software project does vary 'as the product of the number of men
and the number of months', Brooks argues, 'but progress does
not.'[21] Hence, the man-month as a unit for estimating the duration
of a software project is a 'dangerous and deceptive myth', because
'[i]t implies that men and months are interchangeable', when in fact
they are not.[22] To demonstrate this, Brooks points out how the development of a large system 'consists of many tasks, some chained
end-to-end'. When tasks are chained end-to-end, one has to wait for
the conclusion of a certain task before beginning the next one. Thus,
importantly, the time of software development also involves waiting.

Here Brooks does not contest the enlisting of the human and of
time as resources in industrialization. In principle, for him the human and time *are* commodities. What he questions is just the functional equivalence of the human and time as resources *in all parts of
software development*. He claims: '[m]en and months are interchangeable commodities only when a task can be partitioned among many
workers *with no communication among them*'—but this absence of
communication is not a characteristic of software development.[23]
What Brooks is actually trying to demonstrate is that, albeit time
must be quantized, or made discrete, to manage software development, man-time can be useless in certain circumstances—namely,
those circumstances that involve communication. In fact, I want to
point out here that the fallaciousness of Brooks's man-month is the
image of the always imperfect linearization of the time of software
development.[24]

For Brooks, waiting and communication impede the sequencing
of time, and therefore make time estimates unreliable. As for waiting, according to Brooks, '[w]hen a task cannot be partitioned because of sequential constraints, the application of more effort has no
effect on the schedule.'[25] If a project involves non-parallelizable
tasks, its duration cannot be calculated in man-time, because adding
more workforce is useless when tasks cannot be performed inde-
pendently. As for communication, Brooks explains that 'in tasks

that can be partitioned but which require communication among the subtasks, the effort of communication must be added to the amount of work to be done.'[26] Here again man-time cannot be applied because, albeit tasks are parallelizable, if they require communication more man-time is needed to carry out communication itself.[27] Brooks argues that, since software development is inherently a collective effort, 'adding more men . . . lengthens, not shortens, the schedule.'[28] In sum, for him, both waiting and communication make man-time an unreliable unit of measure for time estimates and are obstacles to the linearization and calculation of time. Although insisting on the separation of tasks in order to make the project manageable, Brooks acknowledges that such a separation can never be complete. In fact, he states that when a software project is behind schedule, one adds manpower—and yet, since human beings and time are not interchangeable, 'by adding men, you do not add time.'[29] After adding manpower a project almost always ends up being later than ever. Thus, Brooks proposes his 'Law': 'Adding manpower to a late software project makes it later.' He calls this 'the demythologizing of the man-month'.[30]

Not only does the addition of human power generate more work and make time less and less predictable. Furthermore, as we know, Brooks partially acknowledges the necessity of iteration in the process of software development—that is, the fact that implementation begins before specifications are completed and that code in turn sheds light on specifications' inconsistencies, which eventually leads to specifications being rewritten. It must be noticed here how this iteration calls into question Brooks's conception of 'waiting', and in general of partitioning tasks and of sequencing time. In fact, in order to wait for a certain task to be completed, one must have faith in the possibility of such completion—but the iteration inherent in software development always makes completion uncertain.[31]

In sum, Brooks's 'demystification of the man-month' amounts to the demonstration that the time of software development is never completely calculable. And yet, such calculation—that is, the linearization of time and the attempt to anticipate the duration and outcome of each stage of software development—is necessary in order for software to exist. In Brooks's terms, a perfectly calculable and foreseeable software project would include a finite number of parallel tasks which do not overlap, without iteration and without communication among programmers. According to Brooks himself, such a project cannot exist. Nevertheless, Brooks's strategy for coping with the actual paradoxes of software development involves maintaining the separation between tasks (based on the Platonic distinction between the ideal and the material) and controlling the communication and coordination between them.[32]

This distinction between the ideal and the material is in fact the foundation of Brooks's entire theorization of software development. His model for the organization of the software project is based on the concept of the 'surgical team', which establishes a neat separation between the 'fewer minds' and the 'many hands' involved in any software project.[33] Brooks argues that a software system must be developed at the conceptual level by just one or at most two professionals (the 'minds'), and that other manpower (the 'hands') must be brought in only at the later stage of implementation. In Brooks's model, 'the system is the product of one mind — or at most two, acting *uno animo*.'[34] This sounds very much like Brooks's own dream of the elimination of any need for communication, at least at the 'ideal' level of the project. Furthermore, if we follow Stiegler's conception of the technical exteriorization of consciousness, here we encounter the problem of the multiplicity of consciousness.[35] In fact, never before have we come across a specific discussion of whose consciousness is exteriorized in software. It is Brooks who comes up with the issue of the number of 'minds' at work on a software project. For him, the coordination of the process of software development is not just a matter of making time discrete, nor of making software visible as inscription, but of coordinating a plurality of minds. In fact, if software is exteriorized consciousness, for Brooks perfect software is the exteriorization of only one consciousness.

For the software system to have conceptual integrity, Brooks argues, hierarchy is necessary, as well as 'a sharp distinction . . . between architecture and implementation', while 'the system architect must confine himself scrupulously to architecture.'[36] This means that implementers (the 'hands') must work only on coding without affecting system specifications. At this point, Brooks's famous metaphor of the cathedral enters the picture. '[M]ost programming systems', he muses, 'reflect conceptual disunity far worse than that of [Medieval] cathedrals', whose different parts have been built by different generations of architects and builders. And yet, the disunity of software does not arise from 'a serial succession of master designers, but from the separation of design into many tasks done by many men.'[37]

Brooks's main point here is that 'conceptual integrity is *the* most important consideration in system design' — and he will still contend this twenty years later. A software system needs to 'proceed from one mind'.[38] On the other hand, the pressures of schedule—namely, of the calculated time of software development—require many 'hands', and therefore a division of labour, that must be managed through the clear separation between the specification of the system and its implementation. As in the Garmisch report, specifications are defined as the detailed description of what the system does—a description that institutes the instrumentality of the system, detailing how it can be used as a tool. In terms of the 'what' and the 'how', as we already know, the specifications are the 'what' of the system, while implementation (or the 'realization' of the system) is the 'how'. Brooks illustrates this with the example of the clock, where the specifications correspond to the face, the hands, and the winding knob, from which anyone (even a child) can tell time once they have understood how to read them; the implementation is 'what goes on inside the case'.[39] For Brooks, this separation between specifications and implementation sets in place a kind of 'aristocracy of the intellect', an aristocracy that 'needs no apology', since incorporating ideas coming from implementers would disrupt the conceptual coherence of the system.[40]

Nevertheless, in order to mitigate this exclusion, Brooks adds that implementation is just as creative as specifications writing after all. Implementation simply thrives on the imposition of a limit—consisting of the consolidated specifications of the system—that in turn nurtures the implementers' search for solutions. As long as implementers focus only on the 'how', without ever questioning the 'what', they have the greatest freedom: they are artists that need to be subjected to some kind of containment in order to thrive, but such containment here is not identified with the physical limitations of the computer on which they are working (as one might expect, given Brooks's separation between ideal and material, content and medium), but with the temporal limit established by the division of labour. Thus, the inscription of the software system constitutes *and* limits the freedom of inscription of the implementers. Even more interestingly, while arguing for the creativity—or the relative freedom—of implementers, Brooks presents a more flexible solution to the problem of time sequencing than the one he proposed earlier on in his book. Now he remarks that implementers do not need to wait for the specifications to be completed before starting their job. In fact, as soon as the first draft of the specifications becomes available, they can start implementing 'some functions' and writing 'some code'. Thus, once again, the iteration typical of software development creeps back into Brooks's model. As soon as he allows for an overlap between the activities of specification and coding, he is forced to admit that implementers can make visible some inconsistencies in the specifications of the system. Ultimately, by separating specification from coding, Brooks attempts to perform the impossible expulsion of the 'hand' from the 'mind'—which is precisely what the tradition of originary technicity shows as untenable.[41]

In sum, for Brooks, perfect software is perfectly predictable, and such perfect predictability (or calculation of time) is worked out at the (pure) conceptual level and transmitted (as a pure, invariable content) to a workforce that will embody it into some material substrate. This transmission is a perfect process of communication through noninterfering media. Unfortunately, this perfection has been declared from the beginning as a nonworldly goal. Occasionally, miscommunication can occur: for instance, Brooks establishes a relationship between errors in the software system ('bugs') and miscommunication between 'minds': for him, bugs are the result of miscommunication. But a process of communication that can fail is not perfectly transcendental and transparent. As we already know there is actually no pure content of communication that would be separable from its material substrate.[42] Thus, Brooks's dream of software as the brainchild of oneness reveals itself as an impossible dream.

However, Brooks still attempts to maintain the fragile distinction between specification and implementation through an accurate analysis of the different texts produced in the various stages of software development. He defines specifications as the detailed description of the software system that nevertheless can be modified when inconsistencies are exposed during the phase of implementation. Brooks writes: '[r]ound and round goes [the specifications'] preparation cycle, as feedback from users and implementers shows where the design is awkward to use or build.'[43] In order to control the iteration process, he insists that changes in the specifications 'be quantized' through 'dated versions appearing on a schedule'.[44] Here Brooks starts calling the specifications 'the book', a term that deserves a careful analysis. In fact, 'the book' must have different and clearly separated versions according to the changes introduced in the software system during its development. For Brooks, the history of the different versions of 'the book' is the history of the system. The discretization (or 'quantization') of time here coincides with the consolidated versions of the book, and again writing constitutes the linearization of the time of software development. By establishing all this, Brooks institutes 'the book' as the foundation of the history of the software system—a poignantly biblical image.

Even more importantly, and consistently with the state of the art of memory supports in the 1970s, the book is essentially paper-based. Brooks goes to great lengths to suggest an economical way to introduce changes into the thick and complex 'book'. For instance, he recommends using a loose-leaf folder and retyping just the pages that need to be changed. He also recalls how during the OS/360 project a transition was attempted from a paper-based manual to one based on microfiche technology, and laments that, although the microfiche is much smaller, the user manual remains more manageable *as it can be annotated on the margins*. Brooks's position might seem quite backwards-looking from the contemporary point of view of ever-changing, faster, and smaller digital memory supports. But one must recall here that professionals often take quite a conservative approach to new technologies. Innovation is pursued in terms of software development, but technologies that are considered instrumental to such development (for instance, memory supports for specifications) are treated conservatively.[45] We must also be reminded of Derrida's reflections on the fundamental noninstrumentality of such 'tools'. Famously in *Archive Fever*, Derrida argues that 'what is no longer archived in the same way is no longer lived in the same way.'[46] For him, 'the technical structure of the *archiving* archive also determines the structure of the *archivable* content even in its very coming into existence and in its relationship to the future.'[47] In fact, Brooks's rigid sequencing of time—which is typical of the software engineering of the 1970s—is not separable from the difficulty of erasing and rewriting a written page. One can easily see how paper-based technology shapes the separation between architecture and implementation, because it makes it quite difficult to reinscribe the system when a previous decision is changed. Simply put, *paper shapes software*. Brooks is concerned with maintaining the boundaries between different stages of software development and with containing changes, because the not-so-flexible technology of inscription cannot handle too many reinscriptions of the system (and especially conflicting ones).

Furthermore, Brooks oscillates between the conception of the book as the recording of the history of the software system and as the realization of the system itself. He actually states that the book 'is not so much a separate document as it is a structure imposed on the documents that the project will be producing anyway'—therefore, ultimately, on the system. He writes:

technical prose is almost immortal. If one examines the genealogy of a customer manual for a piece of hardware or software, one
can trace not only the ideas, but also many of the very sentences
and paragraphs back to the first memoranda proposing the product or explaining the first design. For the technical writer, the
paste-pot is as mighty as the pen.[48]

Here Brooks explicitly understands 'the book' as an archive—
namely, as the memory of the changes undergone by the system
during its development. From this point of view—to use Stiegler's
terms—'the book' is a mnemotechnics, since its main objective
seems to be the recording of memory.[49] And yet, the book is also the
first (albeit constantly reworked) inscription of the system—of
which the implementation is but one reinscription. Therefore, the
book *is* the system (technics). Once again, then, not only is Brooks's
distinction between specifications and implementation unmade,
but—as we have already seen in Chapter 3—neither does Stiegler's
distinction between technics and mnemotechnics hold in software.

To recapitulate Brooks's argument in *The Mythical Man-Month*,
the slowness that, for him, characterizes the process of software
development, and that causes missed schedules and a reduced
manageability of projects, is related to the difficulty of coordinating
the different individual consciousnesses that take part in software
development as a process of inscription sequenced in time. Notwithstanding Brooks's attempt to linearize the time of software development and to coordinate it through a supposedly regulated and
transparent communication, the lateness of single software projects
seems to be the reason for a more general slowness in the growth of
software—that is, a broader slowness in technological innovation.

The latter problem is the focus of Brooks's article of 1986, 'No
Silver Bullet—Essence and Accident in SE', where, rather surprisingly, he laments the *absence of technological innovation* in software.
He looks back at the developments of software engineering between
1975 and 1985 and expresses his ultimate scepticism about the possibility of improving productivity in software. There will be no major acceleration in software growth, he prophesizes, since over a
decade there has been 'no single development, in either technology
or management technique, which by itself promises even one order
of magnitude improvement in productivity, in reliability, in simplicity'.[50] The speed of software growth is for him simply *not suffi-
cient* (especially if compared to the fast growth of hardware).[51] In
order to understand Brooks's position better, let me start from the
metaphor of his article's title.

Two familiar themes return in the image of the silver bullet: horror and enchantment. The silver bullet hints at werewolves, which Brooks considers the most terrific of the nightmares that populate Western folklore, because 'they transform unexpectedly from the familiar into horrors.'[52] What terrifies Brooks is the fact that software can turn from something familiar into a monster. He states that 'the familiar software project has something of this character'; it looks 'innocent and straightforward' but it is 'capable of becoming a monster of missed schedules, blown budgets, and flawed products.'[53] Brooks's werewolf is clearly a figure of the unexpected. According to popular wisdom, werewolves can be killed only by silver bullets 'that can magically lay them to rest'.[54] Thus, in Brooks's theory of 1986, software has become a werewolf, a nightmare. Therefore, it must be killed, or be laid to rest. But what magic bullet can lay software to rest? Once again the unexpected takes the form of time spinning out of control (missed schedules) and of the failure to perform according to a plan (flawed products) — which in turn causes the loss of money. Consistently, the silver bullet should 'make software costs drop as rapidly as computer hardware costs do'.[55] However, the silver bullet is a magic weapon, and for Brooks it is bound to remain magic — in other words, unreal. There are no 'startling breakthroughs' in sight concerning the growth of software productivity, and, Brooks remarks, such breakthroughs are 'inconsistent with the nature of software'.[56] 'Startling breakthroughs' are again a figure of the unexpected — but of a very different kind. They are welcome innovations capable of increasing the pace of software growth. Somehow, it could be said that Brooks begs to be surprised by software, but that at the same time he despairs he will never be. He actually thinks that software cannot be surprising enough — that is, that the speed of software production cannot be revolutionarily augmented.

To be fair, Brooks acknowledges that a lot of 'encouraging innovations' are under way. However, these innovations remain at the level of what he calls 'the accidental' (rather than essential) aspects of software — thus bringing back the separation between the ideal and the material that informed *The Mythical Man-Month*. Indeed, for Brooks, it is 'the nature of software' that prevents revolutionary leaps. Software is inherently difficult, while the majority of the past, and even of the current, gains in software productivity deal with the accidental by removing 'artificial barriers' such as 'severe hardware constraints, awkward programming languages, lack of machine time'.[57] Solving these problems does nothing to simplify software's conceptual difficulty which still hampers productivity. Only some revolutionary improvement at the conceptual ('essential') level would make a real difference. In fact, given Brooks's loss of hope with regard to any radical innovation in software development, it might look like software is for him incapable of generating unexpected consequences. But, after a more careful analysis, Brooks's argument offers a different interpretation: what is impossible is not the unexpected as such (since software remains fallible), but a leap of the 'right' kind—that is, a leap that results in a revolutionary increase in productivity, a welcome leap, or an expected 'unexpected'. Ultimately, Brooks hopes for being surprised by something he wishes for—namely, he wants to be surprised without actually being surprised. According to him, what we need in order to speed up software growth is actually *more* control—and the reason why there are no radical innovations in software is that the improvements in control achieved so far are merely of the order of the 'accidental'. Even the most recent innovations in software development (new high-level programming languages, time-sharing, unified programming environments) all remain at the level of the accidental for him. They solve marginal difficulties but do not unlock the essential difficulty of software, which Brooks continues to place at the (mysterious and secluded) level of transcendence.

What I want to argue here is that for Brooks software is predictable—that is, its incapacity for generating revolutionary improvements is predictable—precisely because it is unpredictable, which means fallible. Software will predictably fail (in unpredictable ways), therefore projects will be late, and ultimately software growth will be (predictably) slow. What is most relevant here is that, notwithstanding his lack of hope for any revolutionary changes, Brooks reinscribes unpredictability into software. He reconfirms that software is fallible and asks for more control because he identifies the unexpected consequences of software as failures. If projects are better controlled, software will become less fallible, Brooks believes, and a step-by-step advancement will become possible (all hope for revolutionary advances having been given up). Obviously here Brooks disregards software's capacity for generating different unexpected consequences: for instance, consequences that *are not just failures.*[58]

'Silver' is Brooks's name for 'the unexpected'—namely, the impossible revolution in software. But silver is also what is meant to kill the 'werewolf'—that is, the unexpected as failure. Thus, Brooks wishes for a bullet that would kill the unexpected—but this bullet would have to be unexpected too. Here Brooks confronts the fundamental double valence of the unexpected both as failure and hope. It can be said that, like Derrida's '*pharmakon*', software entails both risk and opportunity, danger, and cure.[59] As I remarked in Chapter 3, the risk and the opportunity implicit in technology cannot be separated. The Garmisch conference report acknowledges that risks are implicit in software, and that software fallibility is unavoidable. But it is precisely when dealing with the issue of the responsibility for the technological risk that the report establishes the impossibility of separating technology from society. Rather than as a disadjustment between the technical and the sociocultural systems, the 'software crisis' at the beginning of software engineering can be understood better within the framework of the mutual co-constitution of the technical and the social. This is the sense of Stanley Gill's ambivalent argument that 'society' should not make 'risky' demands to technology—demands that can be met only by going beyond the current state of technology—while at the same time admitting that technology *always* entails uncalculable risks. In Chapter 3, I argued that the irreconcilability of these two aspects—and therefore the necessity of calculating incalculable risks, and of attributing responsibility for them—is the point where software engineering 'undoes itself' precisely at the moment of its constitution. In fact, I would go as far as to say that, every time we make decisions about technology, we need to take responsibility for uncalculable risks. There is no Habermasian way out of this irreconcilable dilemma: no 'expert' can provide society with all the necessary information to make unrisky decisions. Technology will never be calculable—and yet decisions *must* be made. But risks are also opportunities. Technology is both a werewolf and a silver bullet. Every political decision regarding technology must take into account the werewolf and the silver, the risk and the opportunity—and even the possibility that a risk might *become* an opportunity.

When in *Echographies of Television* Derrida argues that contemporary technology is already in deconstruction because, although programmed and neutralized as controlled 'development', technological innovation still gives rise to unforeseen effects, he might as well have been thinking of the emergence of open source from the software engineering of the 1970s and 1980s.[60] In the rest of this chapter, I will show how, in the open source model of software engineering, Brooks's risks and potential failures become unexpected opportunities for the acceleration of the pace of software growth. It is worth noting at this point that the idea of open source as a more democratic model of software development is quite popular in contemporary digital media studies.[61] Albeit upholding the importance of open access for a political and ethical rethinking of academic knowledge and of the university, in *Digitize this Book!* Gary Hall demystifies many of the dominant discourses surrounding the supposedly unproblematic subversiveness of open source.[62] Indeed, open source can be thought of as a non-hierarchical way of developing software. It actually thrives on the incorporation of 'any' idea — coming from any programmer and user involved in any stage of the process of software development into one version or another of a software product. And yet, I want to emphasize that open source at its beginning was less concerned with the principle of introducing more democracy into software development than with the technical interest in getting things done. Open source relies on a different sequencing of time than commercial software development, but its conception of software is still characterized both by an instrumental understanding of software as well as by the tendency of software to escape its own instrumentality. Ultimately — as a close examination of Eric Steven Raymond's foundational article of open source, 'The Cathedral and the Bazaar', shows quite clearly — the political aspects of the open source model co-emerge with the technical ones, and particularly with the intrinsic instability of the instrumentality of software.

Quite obviously, Raymond's title is, twenty-two years later, a response to Brooks's book of 1975. In his article, Raymond gives an account of his successful open source project, fetchmail, which he started in 1996 and conducted as 'a deliberate test of the surprising theories about software engineering suggested by the history of Linux'.[63] Although Raymond's article is not deliberately presented as a contribution to the field of software engineering, its narrative is actually interspersed with aphorisms about effective strategies for engineering open source software, as well as with explicit evaluations of more traditional models of software engineering. According to Raymond, Linus Torvald developed Linux following a very different methodology from these more traditional models. To explain this point, he contrasts the 'cathedral' and the 'bazaar' models — where the 'cathedral' model is common to most of the commercial world, while the 'bazaar' model belongs to the Linux (and the open source) world. What Raymond calls the 'cathedral' model is in fact software engineering as conceived by Brooks — that is, quite a consolidated discipline with its own established corpus of technical literature. Raymond argues that the two models of the cathedral and the bazaar are based upon contrary assumptions on the nature of software development, and particularly of software debugging.

Software debugging is a late stage of software development, and is part of what in software engineering is generally called 'test phase'.[64] Before being released to commercial users, a software system needs to be tested—namely, it is necessary to verify that the system meets its specifications, or (once again) that it works *as expected*. One of the activities involved in testing is debugging: when a test reveals an anomalous (or unexpected) behaviour of software, code must be inspected in order to find out the origin of the anomaly—namely, the particular piece of code that performs in that unexpected way. Code must then be corrected in order to eliminate the anomaly. The testing process takes time because all the functions of the system need to tested. Furthermore, sometimes the correction of an error introduces further errors or inconsistencies into the system and generates more unexpected behaviour. Although in the phase of testing unexpected behaviour is generally viewed as an error, it is worth noting that decisions must still be made at this level. The testing team is responsible for deciding whether the unexpected behaviour of the system must be considered an error or just something that was not anticipated by the specifications (since, as we have seen earlier on, specifications are never complete) but that does not really contradict them. Errors need to be fixed (by correcting code), but nondangerous (and even useful) anomalies can just be allowed for and included in the specifications. Thus, the activity of deciding *whether an anomaly is an error* introduces changes in the conception of the system, in a sustained process of iteration.[65]

The complexity of the above process explains why software errors are also called 'bugs'. Although the etymology of the term is uncertain, it hints at the fact that errors are often very hard to find—like the moth that Grace Hopper is said to have found trapped in a relay of the electromechanical computer Mark II in 1945, which caused many malfunctions. Locating a bug is hardly a straightforward and unequivocal process. When a malfunction occurs, it is necessary to find out what part of code causes it, and to read it in order to find out what mistake has been made in writing it. Very often no obvious mistakes (such as misspellings) can be found, because the malfunction is the result of the interaction of that piece of code with other pieces of code. Thus, more code has to be inspected, and the process tends to grow exponentially. At this point, Raymond introduces his famous aphorism that 'given enough eyeballs, all bugs are shallow.'[66] This principle is the foundation of the whole conception of open source software engineering. It certainly was the foundation of the Linux project. But in what way is this principle innovative? And in what way could Linux be considered a big leap forward in the history of software development?

In order to answer this question, it is important to notice that, as Raymond states, in 1991 nobody could have anticipated that 'a world-class operating system could coalesce as if by magic out of part-time hacking by several thousand developers scattered all over the planet, connected only by the tenuous strands of the Internet.'[67] He candidly admits that he personally could not foresee it, although he had been involved in open source programming since the early 1980s, being one of the first GNU contributors.[68] Raymond considers Linux an unforeseeable change in the theories and practices of software development, which overturned much of what he thought he knew. He sincerely believed that above a certain critical level of complexity software development required a 'centralized, a priori approach'. He thought that software systems as complex as operating systems needed to be built 'like cathedrals, carefully crafted by individual wizards or small bands of mages working in splendid isolation, with no beta to be released before its time'.[69] To understand this point better, one must be reminded that a beta release (or beta version) is the first version of a software system released outside the group of developers. A beta version is released for the purpose of evaluation on the part of users (who are also called beta-testers). In the commercial world, they are usually prospective customers who receive the software for free and act as free testers. Generally beta releases are not perfect. A beta release is not a prototype—it is a finished system which has been tested and corrected to the point of performing reasonably well. Yet, developers know that it might still present malfunctions: code errors, minor imperfections, unexpected behaviour under very specific circumstances—in a word, problems that are hard to detect in the test laboratory. These problems tend to emerge only when the system is released and starts performing in a 'real' environment—that is, when it is used by nontechnical users in contexts of 'real' complexity that 'stress' all the parts (or functions) of the system. Therefore, the best way to perform the fine-tuning of the system is to release it to beta users, who will interact with it from a 'friendly' perspective, knowing very well that the price they pay for the privilege of interacting with the newest version of the system is that they might encounter unexpected malfunctions. By signaling such malfunctions to programmers, users take an active part in the process of software development. In commercial software, especially in the early 1990s, beta versions are released very cautiously. The official reason is not to 'abuse the patience of the users'. In fact, the absence of beta releases is also a means of controlling the system, a form of secrecy: it prevents the users from seeing the system too early, or from seeing too much of it. By keeping the system hidden from the 'real' users, and by being the only ones who can modify it, 'wizards' believe they protect its consistency.

Linus Torvald's style of development was the opposite of the jealous secrecy of commercial software developers. Raymond describes it in these terms: 'release early and often, delegate everything you can, be open to the point of promiscuity.'[70] Finding out about Torvald's style of development was, for Raymond, a shock. Hence the metaphor: rather than a cathedral, the Linux community resembled 'a great babbling bazaar of differing agendas and approaches'. An example of such bazaars is, for Raymond, the Linux archive site, which 'would take submissions from anyone'.[71] In fact, as we have seen before, Raymond views the fact that a coherent system can emerge from such archives as a kind of magic—or even a miracle: out of the Linux archive, he writes, 'a coherent and stable system could seemingly emerge only by a succession of miracles.'[72] While the cathedral wizards are a figure of secrecy, Linux's magic is a figure of the unexpected. Here magic hints at a revolutionary leap in software growth. And yet, magic also coincides with stability. Raymond's 'miracle' is the 'coalescence' of the system, its stabilization. At the same time, Linux continued to evolve at great speed; it 'seemed to go from strength to strength at a speed barely imaginable to cathedral-builders'.[73] But in what way does this unexpected model of software development reconcile acceleration and the management of time, innovation and stability, software as a process and software as a product—and, ultimately, the instrumentality of software and its capacity for generating the unexpected?

Raymond's description of the fetchmail project, which coincides with the analysis of the process of open source software development, is quite telling. Since 1993, Raymond had been running a free-access Internet service provider (Chester County InterLink, CCL). In 1996, he decided that he needed a new piece of software for CCL, a so-called POP3 for email management. 'So', he recounts, 'I went out on the Internet and found one.'[74] Rather than writing software from scratch, reuse saves time, and a piece of software that is currently being used by other programmers offers guarantees of being well performing. Raymond theorizes reuse in one of his aphorisms:

'Good programmers know what to write. Great ones know what to rewrite (and reuse).'[75] In fact, Torvald did not write Linux from scratch. Rather, he started from a preexisting software system, called Minix, and he decided to improve and enrich it with additional functions. He rewrote parts of it and kept rewriting until the original code of Minix was completely replaced by Torvald's own code. Thus, Minix was overwritten—or, even better, reinscribed—until it became a totally different system—that is, Linux. 'In the same spirit', Raymond comments, 'I went looking for an existing POP utility that was reasonably well coded, to use as a development base.'[76] He picked Carl Harris's software system popclient out of three or four preexisting systems, because it provided 'a better development base' than the others. Thus, he chose popclient because it was somehow easier to extend—in other words, it held a *better promise for the future.*

It is worth remembering here that popclient was already an open source project—that is, a project that had been initiated by Harris but whose code was publicly available to anyone who wanted to reuse it. Raymond describes how Harris, who had lost interest in the program, handed it over to him in the typical etiquette of the open source movement. He comments that '[i]n a software culture that encourages code-sharing, this is a natural way for a project to evolve', and adds the following aphorism: 'If you have the right attitude, interesting problems will find you.'[77] Once again one must be reminded of the early days of software engineering, when software was initially viewed as a problem, but then the problem itself was recollocated (or expelled) in the world 'out there' (in society) in order to define software as a solution. Significantly, the problem that Raymond wants to solve here is itself a piece of software—the CCL system that cannot manage email properly—and the solution can be found in a different 'out there'—namely, 'out there' on the Internet—and it is a piece of software too. Moreover, Raymond understands a problem as something that actively seeks for a programmer with the right attitude to solve it, except that for him the 'problem' seems to never have been anything other than software.

Even more importantly, Raymond claims that he 'inherited' popclient from Harris. What used to be a problem in the corporate world—personnel turnover, viewed with horror by Brooks—in open source is the way in which projects advance. When one is tired of a project, one can pass it on to someone else, and this keeps programmers engaged. Moreover, by inheriting a software system one inherits its user base. Significantly, in open source, users are considered part of the system—actually, its most important part. The precondition for this is, of course, the availability of source code. Raymond observes that users 'can be tremendously useful for shortening debugging time. Given a bit of encouragement, your users will diagnose problems, suggest fixes, and help improve the code far more quickly than you could unaided.'[78]

Here Raymond comes to his most important conclusion: Torvald's real insight was not to underestimate the potential of the users. Therefore, Torvald did not 'invent' Linux. He rather, and more importantly, invented the Linux development model—a new model of software engineering. 'Early and frequent releases', Raymond writes, 'are a critical part of the Linux development model.'[79] Torvald treated his users as co-developers in the most effective way by organizing the time of his project according to the bazaar model: 'Release early. Release often. And listen to your customers.'[80] Around 1991 (the early days of Linux) he sometimes released a new version of the system more than once a day. In sum, according to Raymond, Torvald 'didn't represent any awesome *conceptual* leap forward'; he was not 'an innovative genius' of design, but he was a genius 'of engineering and implementation'.[81] This comment sounds like an implicit response to Brooks's idea that only conceptual innovation could lead to an increase in productivity of software. Torvald did not innovate at design level (which remained quite 'conservative'); rather, he innovated at the level of project organization—a level that Brooks undoubtedly considered 'accidental', and therefore incapable of generating unexpected consequences.

What is important here is that, while apparently deploying the rhetoric of the genius—that is, of the exceptional creative individuality—Raymond is in fact undermining it, because ultimately the realization of an open source project is a collective task. Simply put, Torvald maximized the number of 'person-hours thrown at debugging and development, even at the possible cost of instability in the code and user-base burnout if any serious bug proved intractable.'[82] This passage shows how in open source the maximization of productivity is still the aim—but now programmers are prepared to risk the instability of the system, or rather, they have accepted that instability is the fastest way forward. In a way, it can be said that open source programmers feel comfortable with the idea of working on a Stieglerian device that goes faster than its own time—they even use such speed to manage the project itself. Moreover, they are comfortable with software anomalies, malfunctions, and failures.

Torvald releases different versions of the system very rapidly, because, as Raymond explains, 'given a large enough beta-tester and co-developer base, almost every problem will be characterized quickly and the fix obvious to someone', or, as we have seen above, and according to what he calls 'the Linus's Law': 'given enough eyeballs, all bugs are shallow.'[83]

Thus, we have come full circle to Raymond's initial aphorism. Indeed, Raymond states that, in a conversation with Torvald, he has developed a better formulation of this aphorism—namely, that any error will be sooner or later spotted by someone, and then someone *else* will fix it. The hard part is *finding* it. For Raymond it is quite obvious that 'more users find more bugs', because they all have different ways of stressing the functions of the program (for instance, inventing new uses for it). This effect is amplified when the users are co-developers. It could be said that, in open source, making demands that exceed the boundaries of technology has stopped being a problem. In fact, making unexpected demands towards software seems to be the only way for software itself to grow. And this is not just because users are now empowered with the capacity for 'exteriorizing' or inscribing the system—something that they could also do in traditional software engineering, although there were 'gate-keepers' who 'fended off' users' requests.[84] More importantly, in open source the exteriorization of the software system is explicitly distributed among many individuals, who produce many overlapping reinscriptions of the system where there are no end-to-end tasks. If the reinscription of the system is fast enough, then it coordinates itself. In other words, the potential of speed for managing itself is the unforeseeable consequence of the software engineering of the 1970s and 1980s. Indeed, a totally different point of view on speed (and on software as the exteriorization of consciousness) would have been necessary to foresee that, beyond a certain point, speed would become able to manage itself—or rather, that software would actually overwrite itself quickly enough to be able to coordinate the very process of reinscription. In an open source project, there are no problems of coordination—at least, not of the order of magnitude that preoccupied Brooks in 1995. Programmers do not need to be coordinated, because coordination 'happens' as the fast overlapping of reinscriptions.

However, reinscription involves variation. Every co-developer can potentially spot an unexpected problem in a certain software system and reinscribe the system in a different way. In what way is it then possible to stabilize a system in time? Raymond writes: 'Linus coppers his bets, too. In case there are serious bugs, Linux kernel versions are numbered in such a way that potential users can make a choice either to run the last version designated "stable" or to ride the cutting edge and risk bugs in order to get new features.'[85] Even in open source, software systems need to become stable at certain points in time—that is, any time one wants to stop being a developer and start being a user, one must be able to 'use' the system as a tool. The stabilization of the system coincides with its instrumentalization, or, vice versa, instrumentality emerges with stability in time. And yet, this stabilization is not scheduled; it is not understood as the end of a certain stage (say, specification) and the beginning of another (say, coding). In a way, there are no timetables, no deadlines. Again, stability is something that *happens* to the system, rather than being scheduled and worked towards. However, a certain amount of control needs to be maintained over releases and Linux versions are numbered in order for potential users to choose which version to run. They can either pick a more stable version (which nevertheless might present some anomalies that have not yet been solved) or 'ride the cutting edge' and run a newer version (which is likely to have been further debugged and perhaps also enriched by new functionalities, but which, for this very reason, can give rise to more unexpected consequences). Raymond attributes to users the capacity of evaluating the risks implicit in technology and of minimizing such risks by choosing the more stabilized version of a system. Nevertheless, he has already recognized that software always entails unforeseen consequences, to the point that a system considered stable might actually lead to great surprises. Raymond's distinction between risky and stable systems shows that decisions regarding technology can and must be made taking into account (rather than denying) technology's incalculability.

To conclude, the development of open source out of the more traditional models of software design of the 1970s and 1980s shows how, in software engineering, the instrumentality of software is always (as Derrida would have it) 'in deconstruction', or (in Stiegler's words) it is the unstable, 'pharmacological' result of the process of technological exteriorization. A deconstructive reading of software makes 'originary technicity' most visible and points out the political implications of technological exteriorization. Linus Torvald did not set out to invent a 'more democratic' model for software design; instead, he just wanted to find a way to make Linux work. However, the technical solutions he developed ultimately transformed the modalities of software production and consumption, as well as its distribution and accessibility. Thus, by engaging deeply with technology, Torvald became—to use Raymond's term—an 'accidental revolutionary'. This is why ultimately technology should not be viewed as opposed to political practice (and technical decisions as opposed to political decisions). On the contrary, in the contemporary (and yet originary) co-constitution of the technical and the human, political invention emerges *with* and *through* technology.

NOTES

1. Frederick P. Brooks, *The Mythical Man-Month: Essays on Software Engineering* (Reading, MA: Addison-Wesley, 1995), 32.

2. Eric S. Raymond, "The Cathedral and the Bazaar" (2001a), http://www.unterstein.net/su/docs/CathBaz.pdf. 2; *The Cathedral and the Bazaar: Musings on Linux and Open Source by an Accidental Revolutionary* (Cambridge, MA: O'Reilly, 2001b).

3. One must be reminded here that the very concept of 'operating system' was highly innovative at the time, and that every new operating system substantially innovated on the previous ones. See, for instance, Robert L. Glass, *In the Beginning: Recollections of Software Pioneers* (Hoboken, NJ: John Wiley, 2003).

4. Brooks, *The Mythical Man-Month*, xi.

5. Brooks, *The Mythical Man-Month*, 4.

6. Cf. Peter Naur and Brian Randell, eds., *Software Engineering: Report on a Conference Sponsored by the NATO Science Committee, Garmisch, Germany, 7th to 11th October 1968* (Brussels: NATO Scientific Affairs Division, 1969), 15–16.

7. Brooks, *The Mythical Man-Month*, 4.

8. Cf. Paul Ceruzzi, *Beyond the Limits: Flight Enters the Computer Age* (Cambridge, MA: MIT Press, 1989); Martin Campbell-Kelly, *From Airline Reservations to Sonic the Hedgehog: A History of the Software Industry* (Cambridge, MA and London: MIT Press, 2003).

9. Brooks, *The Mythical Man-Month*, 8.

10. Wendy Chun argues that the 'fetishization' of computer programmes (and particularly of source code) as the product of programmers' wizardry is related to the commodification and industrialization of software (Wendy Hui Kyong Chun, *Programmed Visions: Software and Memory* [Cambridge, MA and London: MIT Press, 2011]). For Chun, such process of fetishization works through the 'solidification' of software, which in turn is the result of the shift of emphasis from software as a process to software as a thing, and from time (the time of software execution) to space (the written form of programmes). Intriguingly, for Brooks a pause in spell chanting equals a space in the text of the program. Although there is no simple equivalence between the use of spacing in a given computer program and the time of its execution, spaces (or blanks) actually determine how a program will be executed. I will propose a detailed analysis of this process in Chapter 5.

11. Naur and Randell, eds., *Software Engineering: Report*, 14.

12. Brooks, *The Mythical Man-Month*, 14.

13. As we have seen in Chapter 3, the Garmisch conference report also gives a relevant role to 'creativity' in software development: it is an unforeseeable

creative leap on the part of programmers that allows the transition between the different stages of software development—particularly, the transition from a supposedly preexistent problem (to which software constitutes the solution) towards software itself (cf. Naur and Randell, eds., *Software Engineering: Report*, 52). As I will clarify in a moment, Brooks proposes a different understanding of 'creativity': for him, creativity constitutes the foundation of the whole process of sequencing in software development.

14. Dorothy L. Sayers, *The Mind of the Maker* (London: Mowbray, 1994).

15. Plato, *The Republic*, 514–519a. Brooks does not reference Aristotle's work directly. However, the distinction between essence and accident is formulated by Aristotle in *Metaphysics* VII 4–6, 10–11 and 17.

16. Brooks, *The Mythical Man-Month*, 15.

17. Brooks, *The Mythical Man-Month*, 15.

18. Naur and Randell, eds., *Software Engineering: Report*, 31.

19. Brooks, *The Mythical Man-Month*, 15.

20. Brooks's expression 'man-month' has obvious sexist undertones. Later on, in 'The Cathedral and the Bazaar', Raymond will propose his own unit of measure, the gender-neutral 'person-hour'.

21. Brooks, *The Mythical Man-Month*, 16.

22. Brooks, *The Mythical Man-Month*, 16.

23. Brooks, *The Mythical Man-Month*, 16, original emphasis.

24. Cf. Jacques Derrida, *Of Grammatology* (Baltimore: The Johns Hopkins University Press, 1976).

25. Brooks, *The Mythical Man-Month* , 17.

26. Brooks, *The Mythical Man-Month*, 17.

27. The fact that the new programmers need training must also be taken into account. Brooks gives the name of 'regenerative effect' to the time spent on training the newly hired programmers, on repartitioning the project ('time re-scheduling') and on communicating in order to coordinate these tasks (Brooks, *The Mythical Man-Month*, 25).

28. Brooks, *The Mythical Man-Month*, 19.

29. Brooks, *The Mythical Man-Month*, 21.

30. Brooks, *The Mythical Man-Month*, 25.

31. The fact that subtasks are actually completed—and that software systems are actually finished and delivered to users—is an effect of time management: at a certain point in time, a system is declared stable (and marketable) and a 'release' is consolidated. Such a consolidation does not prevent system developers from improving and extending the system until they reach a new (and supposedly better) stable version of it. Later on in this chapter I will show open source's different treatment of releases: while the Fordist sequencing of time implicit in Brooks's time management aims at keeping the number of public releases of a system as low as possible (in order to present users with fewer but stable versions of the system), open source software is delivered as often as possible, and users themselves are responsible for debugging it.

32. As we shall see later on, Raymond's strategy will actually downplay the separation between tasks and minimize the control over task coordination—nonetheless maintaining a discrete and linear conception of time.

33. Brooks, *The Mythical Man-Month*, 32.

34. Brooks, *The Mythical Man-Month*, 35.

35. Bernard Stiegler, *Technics and Time, 1: The Fault of Epimetheus* (Stanford, CA: Stanford University Press, 1998), 17–18.

36. Brooks, *The Mythical Man-Month*, 37.

37. Brooks, *The Mythical Man-Month*, 42.

38. Brooks, *The Mythical Man-Month*, 42–44.

39. That Brooks chooses here the very image of the calculation of time as an example is quite appropriate, since—as we have seen in Chapter 3—software instrumentality and the sequencing of time are tightly correlated.

40. Brooks, *The Mythical Man-Month*, 46.

41. Timothy Clark, "Deconstruction and Technology," in *Deconstructions. A User's Guide*, ed. Nicholas Royle (Basingstoke: Palgrave, 2000), 239.

42. Cf. Jacques Derrida, *Limited Inc.* (Evanston, IL: Northwestern University Press, 1988).

43. Brooks, *The Mythical Man-Month*, 62.

44. Brooks, *The Mythical Man-Month*, 62.

45. See, for instance, Jay David Bolter, *Turing's Man: Western Culture in the Computer Age* (London: Duckworth, 1984).

46. Jacques Derrida, *Archive Fever: A Freudian Impression* (Chicago: University of Chicago Press, 1996), 18.

47. Derrida, *Archive Fever*, 17.

48. Brooks, *The Mythical Man-Month*, 75.

49. Bernard Stiegler, *Technics and Time, 3: Cinematic Time and the Question of Malaise* (Stanford, CA: Stanford University Press, 2011), 131.

50. Brooks, *The Mythical Man-Month*, 181.

51. Immediately after its publication, Brooks's article became famous precisely for destroying software engineering's high hopes for increasing the speed of software growth to a level comparable to that of hardware.

52. Brooks, *The Mythical Man-Month*, 180.

53. Brooks, *The Mythical Man-Month*, 180–81.

54. Brooks, *The Mythical Man-Month*, 180.

55. Brooks, *The Mythical Man-Month*, 181.

56. Brooks, *The Mythical Man-Month*, 181.

57. Brooks, *The Mythical Man-Month*, 180.

58. I will show in a moment how the open source movement is exactly one of these consequences.

59. Jacques Derrida, *Dissemination* (Chicago: University of Chicago Press, 1981), 111.

60. Jacques Derrida and Bernard Stiegler, *Echographies of Television: Filmed Interviews* (Cambridge: Polity Press, 2002).

61. See, for example, Matthew Fuller, *Behind the Blip: Essays on the Culture of Software* (New York: Autonomedia, 2003), 30.

62. Gary Hall, *Digitize This Book! The Politics of New Media, or Why We Need Open Access Now* (Minneapolis: University of Minnesota Press, 2008).

63. Raymond, "The Cathedral and the Bazaar," 1. Linux is a Unix-like operating system started in 1991 by Linus Torvald. It is one of the most famous examples of open source software.

64. See, for instance, Ian Sommerville, *Software Engineering* (Boston: Addison-Wesley, 2011). Importantly, Raymond describes software development by using software engineering traditional terms—such as 'specifications', 'coding', 'debugging', 'testing'. These terms, whose emergence I analyzed in Chapter 3, are today widely accepted and deployed in most branches of software development—not just in software engineering.

65. If a bug is incorporated into the specification of the system, it becomes a 'feature' of the system. The distinction between 'bug' and 'feature' is obviously unstable. Even more revealing is the presence, in many software systems, of so-called undocumented features—that is, behaviours of the system which are not covered by a verbal description in the 'user manual' and therefore remain invis-

ible to the user. Software suppliers include undocumented features for many different reasons (for example, for future compatibility) and use them as a way to maintain a certain control on the system (since these 'ghostly' features do not need to be supported and can be removed without warning users). The joke that refers to software bugs as undocumented features ('it is not a bug; it is an undocumented feature!'), which became popular after some of Microsoft's responses to bug reports for its first Word for Windows product, further reveals the instability of the concept of 'bug' (see *Wikipedia*, s.v. 'Undocumented features', http://en.wikipedia.org/wiki/Undocumented_feature). I would like to thank Scott Dexter for pointing this out to me.

66. Raymond, "The Cathedral and the Bazaar," 9.

67. Raymond, "The Cathedral and the Bazaar," 1.

68. GNU is an operating system entirely based on free software. Although being Unix-like, it differs from Unix by being free software; hence its name, which is a recursive acronym for GNU's Not Unix.

69. Raymond, "The Cathedral and the Bazaar," 2.

70. Raymond, "The Cathedral and the Bazaar," 2.

71. Raymond, "The Cathedral and the Bazaar," 2.

72. Raymond, "The Cathedral and the Bazaar," 2.

73. Raymond, "The Cathedral and the Bazaar," 2.

74. Raymond, "The Cathedral and the Bazaar," 3.

75. Raymond, "The Cathedral and the Bazaar," 4.

76. Raymond, "The Cathedral and the Bazaar," 4.

77. Raymond, "The Cathedral and the Bazaar," 5.

78. Raymond, "The Cathedral and the Bazaar," 6.

79. Raymond, "The Cathedral and the Bazaar," 7.

80. Raymond, "The Cathedral and the Bazaar," 8.

81. Raymond, "The Cathedral and the Bazaar," 8, emphasis added.

82. Raymond, "The Cathedral and the Bazaar," 9.

83. Raymond, "The Cathedral and the Bazaar," 9.

84. Naur and Randell, eds., *Software Engineering: Report*, 41.

85. Raymond, "The Cathedral and the Bazaar," 11.

FIVE

Writing the Printed Circuit

For a Genealogy of Code

DO 10 I = 1.10
(FORTRAN line of code said to have caused the crash of the
Mariner I space probe in 1962)

In Chapters 3 and 4, I argued that in the late 1970s and early 1980s, the discipline of software engineering reasserted its capacity for continually opening up and preserving the instrumental conception of software—a capacity that can already be found in the early days of software engineering at the end of the 1960s. I also suggested that the emergence of free/open source programming in the 1990s, with its own brand of software engineering practices and theories, can be understood as one of the unforeseen consequences generated by the conception of software developed in the 1970s and 1980s. In this chapter, I want to take this argument further by showing how software escapes instrumentality not only by generating unforeseen consequences, but also by repeatedly challenging the very process of linearization through which it is constituted. As we have seen, linearization is deeply correlated with instrumentality. In fact, it is precisely by becoming linearized—that is, based on the phonetic alphabet—that writing becomes instrumental to spoken language within the Western philosophical tradition. It can therefore be suggested that linearization constitutes instrumentality—or even that instrumentality is constituted *through* and *as* linearization. And yet, as Jacques Derrida argues in *Of Grammatology*, linearization is nothing but the constitution of the 'line' as a norm—in other words, the line is *only* a model, however privileged.[1] In what follows I intend to demonstrate that such an unstable process of linearization lies at the very basis of the relationship between 'software', 'writing', and 'code' the way they are conceived in software engineering. To do this, I will stray a little from the technical literature of software engineering in its strict sense in order to investigate the development of programming languages in the late 1960s. An analysis of this development will show that programming languages were constituted as linear constructs in order to make software instrumental—or that software has been constituted in the form of the linear sequencing of symbols in order to be thought of as a tool. This chapter can also be considered a conceptual experiment in which I stage an encounter between Derrida's and Stiegler's understanding of technology in order to make sense of what I would tentatively name the 'aporetic' nature of software.

As explained in Chapters 3 and 4, implementation—or the translation of software specifications, however detailed and formalized, into code (that is, computer programs, or texts written in a programming language) is an important stage in the process of software development. In the late 1960s, in the early days of software engineering, quite a large number of high-level programming languages were already in use, and many more were being developed and gradually becoming available—such as FORTRAN, COBOL, PL/1, and Algol. The relative value of some of these languages was actually a topic of discussion at the two NATO Conferences on Software Engineering.[2] Nevertheless, and even more importantly, the foundation of *the theory of programming languages*—or the general theorization of what a programming language ought to be—was being laid out systematically only at the very end of the 1960s. In what follows I will look at one of the foundational texts of the theory of programming languages—namely, *Formal Languages and Their Relation to Automata*, published by John E. Hopcroft and Jeffrey D. Ullman in 1969 (the year of the second NATO Conference on Software Engineering)—in order to investigate to what extent the process of 'linearization' can be shown to be at work in the constitution of the concepts of 'programming languages', 'software', and 'code'. I want to focus on how professionals and academics struggled with the idea of linearization in order to make software manageable.

Since linearization and instrumentalization go hand in hand in the theory of programming languages, a deconstructive reading of Hopcroft and Ullman's text will be able to problematize them together. In this chapter, I am also taking on Stiegler's understanding of *différance* as the possibility of studying the differentiation of technical objects empirically and historically in order to show how software is constituted in the theory of programming languages as an *aporetic* concept. As we shall see, software constantly escapes the binary conceptual distinctions that define it (language/code, software/hardware, source code/execution, program/process, ideal/material). These distinctions can only be accessed in their singularity — that is, as they are configured at certain moments in the history of software (for example, in the texts of the theory of programming languages). Stiegler's distinction between technics and mnemotechnics must be completely revised here, precisely in order to give an account of the differentiation of what Stiegler would call the 'organized inorganic matter' of software.[3] If read deconstructively, the theory of programming languages shows how software differentiates itself as a technical object through a process of constant reinscription. This process of differentiation can be understood without a recourse to the distinction between technics and mnenmotechnics — that is, to the idea of the 'aim' of software (as a recording or non-recording technology). More importantly, the constant self-differentiation of software — that is, its continuous reinscription — entails what earlier on I called the 'unpredictability' of software. In this unstoppable process of self-differentiation, some of the reinscriptions of software cannot be anticipated from the point of view of what Stiegler would call 'the *who*' that takes part in the process (that is, from the point of view of programmers or users). In fact, I want to argue that the moment in which one realizes that software is not functioning as expected, and decides to intervene on such unexpected functioning, corresponds to the moment in which 'the *who*' emerges from and through its interaction with 'the *what*' (software) — or the human emerges from and through its interaction with the technical. But let me now give a concrete example of this process by offering a close examination of Hopcroft and Ullman's work. Whenever Hopcroft and Ullman's account is so synthetic that it risks sounding unclear, I will also refer to some later texts, and especially to the handbook on compilers coauthored in 2007 by Alfred V. Aho, Monica S. Lam, Ravi Sethi, and Jeffrey D. Ullman, in order to clarify concepts.[4]

The late 1960s were not only the time of the first two conferences on software engineering; they were also the time of substantial changes in the development of programming languages. By 1968, there existed a vast number of high-level programming languages, albeit only about fifteen were in widespread use, the most common being FORTRAN and COBOL. To put this in context, one must bear in mind that the first digital computers appeared in the 1940s and were used for scientific applications. As Robert W. Sebesta states in his brief history of programming languages, '[t]he early high-level programming languages invented for scientific applications were designed to provide for those needs', which throughout the 1950s and 1960s basically coincided with the need for efficient floating-point arithmetic computation.[5] However, the use of high-level languages was not yet consolidated in all areas of computing, as the debate at the NATO Conferences on Software Engineering showed. In fact, the other main area of interest at the time of the NATO conferences was related to the commercialization of computers, and thus to 'systems programming'. One of the most significant issues of the time involved having to decide whether operating systems and all the programming support tools of a computer system, collectively known as *systems software*, which were still being provided by manufacturers for free, should be priced and commercialized separately. This debate was also concerned with what languages were best suited for systems programming. Being used almost continuously, systems software needed to be efficient; therefore, it had to be written in programming languages that could be executed quickly. Furthermore, such languages had to provide 'low-level features'—or instructions that allowed programmers to control the machine's physical architecture more directly, without the abstraction which is otherwise a crucial advantage of high-level languages. An important controversy at the NATO conferences was whether it was possible to use high-level languages for systems programming.[6] Many participants in the NATO conferences deemed it impossible; they actually claimed that they preferred languages with 'low-level features' because they wanted to keep 'close to the machine'. As I explained in Chapter 3, the distinction between programming languages and 'the machine', or between language and materiality, was already quite unstable; it was, so to speak, 'in the process of making and unmaking'. The general anxiety about losing touch with materiality could be interpreted as the fear of losing the ability to influence the functioning of computer memory and circuits. However, in the 1960s and 1970s, the 'software scene' (to use Randell's term) saw the development of special machine-oriented high-level languages. These languages were meant to be used for writing systems software for different machines produced by different computer manufacturers (for example, PL/S was developed to write systems software for IBM mainframe computers). Finally, it is worth remembering that the issue of programming languages concerned not just systems software but also real-time applications, such as military applications, which were most likely at the core of NATO's interest in programming languages.

In sum, high-level languages were a reality at the end of the 1960s. Hopcroft and Ullman published their work on languages and automata in 1969. Although the theory of formal languages was already a dynamic subfield of computer science, this book took it one (substantial) step further. It constituted a systematization of the authors' research in the field, previously disseminated in journals of mathematics and computer science and in notes for courses on the theory of formal languages that the authors had taught at Princeton, Columbia, and Cornell University. The aim of the book was to present 'the theory of formal languages as a coherent theory' and to make explicit its relationship to automata.[7] It represented the most synthetic account of the theory of formal languages in 1969—a theory that software engineering relied on (and took for granted) when conceptualizing the stage of 'implementation' in the process of software development.

The theory of formal languages sprang to life in the mid-1950s, when Noam Chomsky developed a mathematical model of what he called a 'grammar' in connection with his study of natural languages.[8] In fact, the most powerful and productive influence of his theory is to be found in the field of programming languages, since the formal grammars he developed (also known as 'Chomsky's grammars') ended up providing the foundation for the design of programming languages. In Hopcroft and Ullman's book, the Chomskyan concept of 'grammar' is established as the fundamental descriptive device for formal languages.[9] It must be emphasized here that Hopcroft and Ullman consider the pre-Chomskyan definitions of language (for instance, Webster's definition of language as 'the body of words and methods of combining words used and understood by a considerable community') as not sufficiently precise for computing. The goal of Hopcroft and Ullman's book is thus to 'define a formal language abstractly as a mathematical system'.[10] In other words, in order to be useful in computing, a language needs to be abstract, or formalized according to a mathematical model.

This association between formalization and instrumentality goes back at least to Husserl's reflections on geometry. As we have seen in Chapter 1, in *Technics and Time* Stiegler shows how Husserl associates algebra, or the formalization of geometry, with instrumentality. Algebra, as a technique of calculation, is nothing but a formalism—and, as any formalism, it is devalued as 'eidetically blind', instrumental, or 'just technical'.[11] From this point of view, Hopcroft and Ullman's attempt to formalize language—or, better, to construct a formal theory of languages—is also an attempt to make language instrumental. Formal languages (and therefore also programming languages), it can be said, are constituted *as* instrumental. However, as we shall see in the remaining part of this chapter, this formalization is simultaneously established and undone.

Hopcroft and Ullman begin their analysis of formal languages by providing the following definition of an alphabet: '[a]n alphabet or vocabulary is any finite set of symbols.' They explain: '[a] set is countably infinite if it is in one-to-one correspondence with the integers (i.e., if it makes sense to talk about the *i*th element of the set).'[12] Thus, alphabets are *finite* sets of symbols which

> include digits, the Latin and Greek letters both upper and lower case (possibly with combinations of subscripts, superscripts, underscores, etc.), and special symbols such as #, ¢, and so on. Any countable number of additional symbols that the reader finds convenient may be added. Some examples of alphabets are the Latin alphabet, {A, B, C, …, Z}, the Greek alphabet, {α, β, γ, …, ω}, and the binary alphabet {0, 1}.[13]

The first important aspect to be noticed here is that this is a *written* alphabet, since it is case-sensitive—it distinguishes between upper and lower case—and it includes a number of special symbols with no phonetic equivalent. It is also clearly a Western-centric one, because it includes Greek and Latin characters. Even more importantly, Hopcroft and Ullman give examples of what they call 'natural' alphabets—which I want to rename here as alphabets associated with 'natural' language. As we have seen in Chapter 1, if we follow Derrida's rereading of Leroi-Gourhan in *Of Grammatology*, no 'natural' alphabet actually exists: any alphabet is always already technological.[14] The concept of the alphabet—including alphabets associated with natural languages—always entails a form of linearization. Furthermore, it is worth noting that, along with the 'natural' alphabets, Hopcroft and Ullman introduce the notion of the 'binary alphabet'. In this extremely relevant passage, binary notation (which is the foundation of digitization) is presented as just another alphabet, no more and no less (un)natural than all the other possible alphabets, and containing just two symbols. In sum, here digitization is presented as an alphabet—that is, as a process of linearization.

Having defined the alphabet, Hopcroft and Ullman proceed to define the 'word'. They write: 'A *sentence* over an alphabet is any string of finite length composed of symbols from the alphabet. Synonyms for sentence are *string* and *word*.'[15] 'String' is one of the most commonly used terms in the theory of programming languages. A string is a finite, ordered sequence of discrete symbols drawn from the alphabet.[16] The concept of the string makes it even clearer how the theory of formal languages works as a process of *linearization* (especially if we consider that a computer programme can also be defined as a string, as we shall see later on). For instance, if we consider the binary alphabet (which Hopcroft and Ullman define as {0, 1}), an example of a string is: 00110110001101. One has to bear in mind that the binary alphabet is a very small alphabet with which an infinite number of strings can be produced. An important point that can be drawn out of this definition of the binary alphabet is that computation minimizes alphabetic requirements. Significantly, for Hopcroft and Ullman, natural and programming languages are not as different as they are for Hayles.[17] Hopcroft and Ullman place all languages in a sort of continuum, where it is possible to define an appropriate alphabet for each language and to construct the sentences (or strings of symbols) that belong to that language. For them, the difference between 'natural' and programming languages seems to reside in the different kinds of 'grammars' that are necessary to 'describe' these languages. The continuum between natural and programming languages is based on linearization as well as on the technological nature of the alphabet.

In sum, we have ascertained so far that a string 'over an alphabet' can be any finite sequence or string of letters belonging to that alphabet, and that a 'language' is any discrete set of strings 'over' some fixed alphabet.[18] It is worth giving some consideration to the word 'over', which Hopcroft and Ullman use regularly. The word 'over' defines the alphabet as a space, which we could call the space of linearization. A language can be defined only 'over' an alphabet—in other words, somehow 'within' it and 'in terms of' it. In order to have a string we must have an already linearized system of reference—that is, the string presupposes the linearization that has always already occurred in the alphabet, which in turn can produce repetitions of discrete elements. Tellingly, Hopcroft and Ullman eventually define language as follows: '[a] language is any set of sentences over an alphabet. Most languages of interest will contain

an infinite number of sentences.'[19] Within the linearized space of
the alphabet the possibility of infinity is allowed: symbols can be
infinitely recombined, but only in linear terms, as sequences. Moreover, the length of a string is finite (there is no infinite string), but
there can be an infinite number of strings, or infinite linear recombi-
nations of discrete symbols. This is a kind of infinity which is kept
within the limits of linearization.

At this point, Hopcroft and Ullman raise the important question
of how to 'represent (that is, specify the sentences of) a language'.[20]
They write:

> If the language contains only a finite number of sentences, the
> answer is easy. One simply lists the finite set of sentences. On the
> other hand, if the language is infinite, we are faced with the
> problem of finding a finite representation for the language. The
> finite representation will itself usually be a string of symbols over
> some alphabet, together with some understood interpretation
> which associates a particular representation to a given lan-
> guage.[21]

Simply put, the 'representation' of a language is defined in terms
of the 'specification' of its sentences. In this way, representing a
language amounts to *writing down* all its sentences. This kind of
'representation' is therefore an *inscription*. In an analogous way, as
shown in Chapter 3, the participants in the NATO Conferences on
Software Engineering thought that specifying a software system
amounted to repeatedly reinscribing it—that is, writing and rewriting a description of it until this description 'became' the system
itself. However, as Hopcroft and Ullman remark, listing all the sentences of a language works well for finite languages: in other words,
languages that have a finite number of sentences can be linearized
in the form of a list. But infinity raises a problem. Infinite languages
require a different kind of inscription, which must be capable of
dealing with infinity while remaining itself finite (in fact, representation can also be considered a string, and a string is finite by definition). However, for Hopcroft and Ullman, the representation of an
infinite language is a very special kind of string, which they call a
'grammar'. In other words, a grammar is just another material inscription of what the authors call 'language'. But what are the characteristics of this particular inscription they call 'grammar'?

To give but a simplified synthesis of Hopcroft and Ullman's argument, a grammar is a set of rules to which the sentences of a
given language must conform in order to be part of the language.

Hopcroft and Ullman aim at finding a way to embody such a set of rules in an automatic and finite sequence of steps capable of defining the boundaries of a language by deciding which sentences are correct and by ruling out the incorrect ones. Such a finite sequence of steps could well be carried out by a computer program. In fact, we shall see that there are different kinds of computer programmes whose purpose is to analyze formal languages. Moreover, since a computer programme is ultimately a string of symbols, the strings that these computer programmes analyze can themselves be computer programmes. A particular category of grammar-aware programmes that carry out the analysis of other computer programmes are called 'compilers'. In fact, compilers are specifically tasked with 'translating' programmes from one language to another. I am going to examine the functioning of compilers in greater detail in the further section of this chapter. For now, it is worth noting how, being computer programmes, compilers are themselves written in programming languages, which in turn have been specified and (typically) compiled. Thus, in principle it takes a compiler to generate a compiler. [22]

Let me now examine in some detail Hopcroft and Ullman's discussion of the concept of grammar. They explicitly refer to Chomsky's article of 1956 entitled 'Three Models for the Description of Language'. [23] As we have seen, in the non-phonetic realm of programming languages linearization functions as the discretization and sequentialization of computation into alphabets and grammars—a linearization whose time is *not related to sound* (it is not the time of the pronunciation of speech). Rather, it is the time of software compilation and execution by the computer (and hence it is ultimately regulated by the computer's internal clock). However, to explain the functioning of grammars better, Hopcroft and Ullman choose to analyze an English sentence as a way to clarify the basic concepts of the 'tree-diagram' and of the 'rules of production'. They write:

> For example, the sentence 'The little boy ran quickly' is parsed by noting that the sentence consists of the noun phrase 'The little boy' followed by the verb phrase 'ran quickly'. The noun phrase is then broken down into the singular noun 'boy' modified by the two adjectives 'The' and 'little'. The verb phrase is broken down into the singular verb 'ran' modified by the adverb 'quickly'. . . . We recognize the sentence structure as being grammatically correct. [24]

Hopcroft and Ullman start out by 'parsing' the given sentence. 'Parsing' is a term that has become common in the theory of programming languages and of compilers, and it broadly means 'breaking down into identified components'. The sentence structure is then diagrammed in the form of a tree-diagram. Hopcroft and Ullman's process of parsing is meant to prove the grammatical correctness of the analyzed sentence—which is obviously divorced from the question of whether this sentence has any meaning. The process of parsing, being a process of 'breaking down' a given string, consists in looking for blanks or word separators (and it must be kept in mind that, when viewing the sentence as one string, the blanks are just part of that string, or symbols in the alphabet) and in decomposing the string into substrings according to certain rules. Hopcroft and Ullman explain:

> The rules we applied to parsing the above sentence can be written in the following form:
>
> <sentence> → <noun phrase> <verb phrase>
> <noun phrase> → <adjective> <noun phrase>
> <noun phrase> → <adjective> <singular noun>
> <verb phrase> → <singular verb> <adverb>
> <adjective> → The
> <adjective> → little
> <singular noun> → boy
> <singular verb> → ran
> <adverb> → quickly
>
> The arrow in the above rules indicates that the item to the left of the arrow can generate the items to the right of the arrow.[25]

In this (Chomskyan) formalism the names of the parts of the given sentence (such as 'noun', 'verb', 'verb phrase', etc.) are enclosed in brackets. The rules listed by Hopcroft and Ullman appear as rules of replacement (or substitution) of one string with another: the process is explained as a reinscription of some bracketed string into one (or more) bracketed string or one (or more) non-bracketed string. This reinscription of strings is governed by the arrow (the typical operator of the 'rules of production'). The arrow indicates a movement by which a certain inscription is replaced by another one within the discrete and ordered dimension of the line.

As Hopcroft and Ullman explain, the strings enclosed in angular brackets, such as <singular noun>, <verb phrase>, <sentence>, from which strings of words can be derived, are called 'syntactic categories', 'nonterminals', or 'variables'. The strings which are not enclosed in angular brackets—that is, as the authors explain with characteristic awkwardness caused by the self-reflexivity of their meta-language, 'the objects which play the role of words'—are called 'terminals'. The relations that exist between strings of variables and terminals (that are indicated by the arrow) are called 'productions'. Examples of productions are <noun phrase> → <adjective> <noun phrase> or <singular noun> <singular predicate> → <singular noun> <adverb> <singular verb>. A production that has the form <noun phrase> → <adjective> <noun phrase> permits the generation of an infinite number of sentences because it allows the syntactic category <noun phrase> as a possible substitution for itself. Such a rule is said to be 'recursive'. Recursivity makes it possible to generate (and therefore, to specify) an infinite number of strings through a finite number of rules. This is exactly the goal that Hopcroft and Ullman set themselves at the beginning—to find a finite specification for an infinite language. I want to point out here that recursivity allows infinity to emerge from finity—in other words, one can keep applying recursivity indefinitely, thus generating infinity. However, Hopcroft and Ullman remark once more that, although an infinite number of sentences can be generated by applying some rules indefinitely, any single string must stop at some point, therefore being finite. Actually, they speak here of strings as 'quantities', thus clearly showing that a formal grammar is intended to work on a computer (which can only deal with finite quantities).

In order to 'formalize the notion of grammar', Hopcroft and Ullman propose a specific notation for grammars. Besides the definitions of nonterminals, terminals, and productions, another definition is given, that of the 'start symbol'—which is one nonterminal that 'generates exactly those strings of terminals that are deemed in the language'.[26] There is only one start symbol per language. In Hopcroft and Ullman's example, <sentence> is the start symbol. As Aho et al. point out, '[u]nless stated otherwise, the head of the first production is the start symbol.'[27] It can be said that, with the start symbol, a point of origin is established for the language. Actually, a language consists of all the strings of terminal symbols that can be generated *by* the start symbol and a particular grammar.

Finally, Hopcroft and Ullman provide the following formal *notation* for a grammar: 'we denote a *grammar* G by (V_N, V_T, P, S). The symbols V_N, V_T, P, and S are, respectively, the *variables, terminals, productions,* and *start symbol*. V_N, V_T, and P are finite sets.' Furthermore, 'the set of productions P consists of expressions of the form $\alpha \rightarrow \beta$, where α is a string in V^+ and β is a string in V^*. Finally, S is always a symbol in V_N.'[28] This definition makes very clear once again the written nature of grammars. In fact, terminal symbols are lowercase letters from the beginning of the alphabet (such as *a*, *b*, *c*); operator symbols such as +, *, and so on; punctuation symbols such as parentheses, commas, and so on; digits from 0 to 9, and (as a later text will add with specific reference to the generation of programming languages) 'boldface strings such as **id** or **if**, each of which represents a single terminal symbol'.[29] Nonterminals are uppercase letters which occur early in the alphabet (such as A, B, C), the letter S (which, when used, is usually the start symbol), and 'lowercase, italic names such as *expr* or *stmt'*.[30] Besides, and even more importantly, the process of replacement described by the rules of production, and by the branching out of the tree-diagram, is presented here not only as an orientated linearized movement, but also as a movement from left to right that follows the Western movement of the eyes of the reader—a reader, that is, of alphabetical writing. This has even more interesting implications when we consider the tree-diagram. In a subsequent passage, Hopcroft and Ullman define what they call the 'derivation tree for context-free grammars' as follows:

> We now consider a visual method of describing any derivation in a context-free grammar. A *tree* is a finite set of *nodes* connected by directed *edges*, which satisfy the following three conditions (if an edge is directed from node 1 to node 2, we say the edge *leaves* node 1 and *enters* node 2):
>
> 1. There is exactly one node which no edge enters. This node is called the root.
> 2. For each node in the tree there exists a sequence of directed edges from the root to the node. Thus the tree is connected.
> 3. Exactly one edge enters every node except the root. As a consequence, there are no loops in the tree.[31]

This visual representation of linearity in/as a tree deserves careful analysis. Movement and time are also part of the tree. There are departures and arrivals: edges have one and only one direction; every edge describes one orientated movement; an edge 'leaves' one 'node' and 'enters' another. It must be pointed out that the tree reaffirms, first and foremost, the existence of a point of origin. The origin is defined by its being unoriginated, and visually by its impenetrability: no edges enter it, and it is the only unentered node. The process of linearization here establishes a point of origin for itself. Such a point of origin is nominated as a 'root'—where the metaphors of kinship and genealogy actually seem predominant over the metaphors of biology, since the tree grows upside down, as Jay Bolter once noticed in relation to the many trees of computer science.[32] Importantly, all the nodes are connected to the root. There is no stand-alone node in the tree; moreover, there is no node unconnected to the origin. Although the edges are orientated from departure to arrival, every node is traceable backwards; all nodes have a common origin. We can always go backwards; time can always be reverted to its starting point. To make traceability possible, the movement throughout the tree needs to be controlled. Therefore, the third point listed by the authors is also very important, since it excludes loops from the linearized space of the tree. Not only is the tree linear; it is also loop-free. This is achieved by handling nodes so that every node allows only one edge to enter it, with the origin still being the exception. In fact, while reversals, or returns to the origin, are possible, what is impossible is entering a reiteration that causes infinite loops. In sum, a certain kind of repetition is formally expelled from the tree. However, as we already know from Derrida, it is the possibility of repetition that constitutes the sign as material trace, and thus also the possibility of symbols, alphabets, and trees.[33] Moreover, as recursivity is to a certain extent equivalent to repetition, the recursive nature of the tree makes it intrinsically repetitive. Similarly, as shown in Chapters 3 and 4, the loop constitutes a problem in software design, while at the same time making software design possible: reiteration (the repeated attempt at constructing the software system) is both what makes the system possible and what threatens it with the impossibility of ever being finished, or with producing unexpected consequences as unforeseeable variations through repetition. In a sense, iteration always appears to be unmanageable in software at each level, from the tree-diagram of context-free grammars to the broader process of software development. However, since iteration is also unavoidable, its expulsion from software is impossible. In fact, iteration—that is, the process of generating differences in repetition—is a constitutive characteristic of software, just as fallibility (and the capacity for generating unforeseen consequences) is constitutive of technology. The expulsion of the loop from the tree corresponds to the impossible expulsion of risk from technology. In fact, the definition of iteration as difference in time makes reinscription always risky, because it is always open to variation. I will return to this important point in a moment. For now, let me explore the relationship between the string and the tree a little further, in order to show how the process of reinscription links the two. Hopcroft and Ullman give the following definition:

> The set of all nodes *n*, such that there is an edge leaving a given node *m* and entering *n*, is called the set of *direct descendants* of *m*. A node *n* is called a descendant of node *m* if there is a sequence of nodes $n_1, n_2, \ldots, n_k$ such that $n_k = n$, $n_1 = m$, and for each *i*, n_{i+1} is a direct descendant of n_i. We shall, by convention, say that a node is a descendant of itself. [34]

Although here the term 'descendants' makes the kinship metaphor even more clear, the most important point in this passage is the direction of the edges. Such direction, in turn, determines how a given node is classified and how the reinscription of the tree as a string is performed. Hopcroft and Ullman continue:

> Some of the nodes in any tree have no descendants. These nodes we shall call *leaves*. Given any two leaves, one is to the left of the other, and it is easy to tell which is which. Simply backtrack along the edges of the tree, toward the root, from each of the two leaves, until the first node of which both leaves are descendants is found. If we read the labels of the leaves from left to right, we have a sentential form. We call this string the *result* of the derivation tree. [35]

This description of the passage from tree to string is extremely significant. This passage from the two-dimensional space of the tree to the one-dimensional string is realized through an orientated movement either from the left to the right or from the right to the left. The space of the tree is therefore an orientated space. The most important aspect here is that it is possible to distinguish the left from the right *because* the left and the right actually coincide with the left and the right of the written page (that is, the left and the right related to print, to the book, to traditional Western writing). Thus, the passage from the tree-diagram to the string that is characteristic of formal grammars can only exist *because* there is a consolidated tradition that establishes what a page is, what the left and the right of the page are, what it means to read, and in what direction one reads. In this sense formal grammars, the programming languages which are based on formal grammars, and programs written in those programming languages can all exist *because* of the page. [36]

Even more importantly, however, the reinscription of the strings of a language as rules of production or as a tree-diagram is made possible by the presence of the graphic interruption within strings. In Hopcroft and Ullman's example, the interruption, understood as the separation between English words, is what enables both the transcription of the given English string as a set of rules of production, and the transcription of the rules of production as a set of English strings. The same can be said for programming languages. For instance:

stmt → **if** (*expr*) *stmt* **else** *stmt*

The above rule of production establishes that an if-else statement in the Java programming language is the concatenation of the keyword **if**, an opening parenthesis, an expression, a closing parenthesis, a statement, the keyword **else** and another statement. In this rule of production, **if** and **else** are terminals, so are the parentheses, while *stmt* (statement) and *expr* (expression) are nonterminals (defined by other rules of production). The above rule of production is expressed using the Backus Naur Form.[37]

In order to understand better the role played by the symbols of interruption (such as the semicolon or the blank) in programming languages, it is worth now examining in detail the concept of the 'compiler'. Although the main body of Hopcroft and Ullman's book is devoted to the relation between grammars and automata, and although there is a relationship between the concept of automata and the concept of a compiler, Hopcroft and Ullman do not give a discursive account of what a compiler is. At the cost of some oversimplification, it can be said that a compiler is a program that uses the logic of specific classes of automata (namely, DFA—or Deterministic Finite Automata—and pushdown automata) to parse computer programmes (or 'source code'). Let me now take a little detour from Hopcroft and Ullman's book in order to explain how compilers work and why they are crucial in my argument.

During the phase of the implementation of a software system, after programs have been written the implementation itself is not really finished. What has been written is generally a large ensemble of interconnected programs, expressed in some high-level programming language. All this, as argued in Chapters 3 and 4, forms the state of the software system at this point. For instance (assuming that the system is written in C language), a very tiny program fragment might have the following form:

```
int main()
{
int x, y, z;
x=x+y;
return 0;
}
```

This example makes clear that a program is inscribed in a form that takes into account the space of the page (even if the page is visualized on a screen) and that respects the conventional direction of (Western) reading. However, from the point of view of the theory of formal languages, the program is one continuous string, with blanks and semicolons coded as ASCII characters that are also part of the string. In fact, the use of new lines and tabulations has the only function of making the program more easily readable for the human eye. In order to be reinscribed *as* circuits (another *state* of software), this program needs to undergo not just one but many reinscriptions. When I say that the program is going to be reinscribed as circuits, I do not mean to argue that 'there is no hardware'—to paraphrase Kittler's paradoxical argument that 'there is no software.'[38] On the contrary, I am trying to emphasize how the (unstable) distinction between what are generally called 'hardware' and 'software' emerges from the process of internal differentiation of software as material inscription. In a way, it could be said that there is no concept of software that does not constitute its concept of hardware—and vice versa.

Aho et al. remark that '[b]efore a program can run, it first must be translated into a form in which it can be executed by a computer. The software systems that do this translation are called compilers.'[39] For the sake of clarity, let me now follow Aho et al. in their informal and intuitive explanation of compilers. I will come back to Hopcroft and Ullman's text at the end of this section. Also, while I am examining here the description of a compiler given by Aho et al., it must be kept in mind that I am not focusing on the actual functioning of the compiler, but rather on the process of reinscription undergone by the originary program (the source program), in order to show how such process works. Aho et al. define a compiler as follows:

> Simply stated, a compiler is a program that can read a program in one language—the source language—and translate it into an equivalent program in another language—the target language. . . . An important role of the compiler is to report any

errors in the source program that it detects during the translation process.[40]

A compiler is thus, to use Hopcroft and Ullman's term, an automaton that 'reads' any program previously written in the high-level programming language for which the compiler has been designed (the 'source language') and reinscribes it—usually—as binary code (the 'target language').[41] The terms 'reads' and 'translates' used in the above passage are metaphors for the material reinscription of the source program—a process of reinscription that is technically called 'compilation'. The process of compilation can be broken down into different phases. Aho et al. write:

> The first phase of a compiler is called *lexical analysis* or *scanning*. The lexical analyzer reads the stream of characters making up the source program and groups the characters into meaningful sequences called lexemes. For each lexeme, the lexical analyzer produces as output a token . . . that it passes on to the subsequent phase, syntax analysis.[42]

The process of compilation consists first of all of detecting 'interruptions' in the string of the program (such as blanks and semicolons) and of regrouping the symbols of the string into smaller substrings called 'lexemes'. It must be noted that the compiler is an automaton that follows the rules of production of a specific grammar—and it is this grammar that establishes what works as an interruption in the source program (for instance, a blank space or a ';' work as interruptions, but not an 'a').[43] The symbol of interruption therefore ruptures the chain of symbols and functions as a discriminating tool for re-grouping the other alphabetical symbols into new groups ('lexemes'). For each lexeme, the phase of lexical analysis produces a 'token'—in other words, each lexeme is reinscribed as a slightly different string with a few explanatory symbols added.[44] At this point, the initial single string of the source program has been completely reinscribed. Thus, the lexical analyzer also works on the basis of linearization and discreteness: its function is to locate the ruptures in the source program, to make them work *as* ruptures, to make them function in the context of the computer so that the originary string is transformed into a different string (the sequence of tokens). The lexical analyzer works with the rupture in order to reinscribe the string. The ruptures (that is, the symbols without semantical significance which serve the purpose of breaking down the string) disappear from the initial string as symbols (that is, they disappear as characters with their own binary encoding in the com-

puter memory) and they are reinscribed literally *as* the 'tokenization' of the program. They are actually what enables such tokenization. Again, it must be noticed that no single agent can be identified
here that 'originates' this process of reinscription (neither the compiler, nor the programmers who wrote the compiler and/or the compiled string; not even the string itself). Rather, reinscription is articulated as a self-differentiating process of material transformation.

Starting from the sequence of tokens produced in the phase of
lexical analysis, another reinscription is carried out. Aho et al. continue: 'The second phase of the compiler is *syntax analysis* or *parsing*.
The parser uses the first components of the tokens produced by the
lexical analyzer to create a tree-like intermediate representation that
depicts the grammatical structure of the token stream.'[45] Again, it is
not important to understand all the technicalities of this passage.
What is important is that the tokens are 'rearranged' once more
according to the syntactical rules of the grammar on which the language is based, with the help of the additional symbols inscribed
into the tokens in the previous phase. The way this happens varies
physically: obviously no graphical representation of a tree is actually 'depicted' in the memory of the computer. The result of the parsing process is simply another linear inscription of the source code.
In other words, the sequence of tokens is regrouped in order to be
then analyzed by the 'semantic analyzer', which in turn checks the
consistency between the tokens and produces the so-called 'intermediate code'. The authors explain: 'In the process of translating a
source program into target code, a compiler may construct one or
more intermediate representations, which can have a variety of
forms.'[46] They define the intermediate code as 'assembly-like', and
we shall see in a moment what such intermediate code looks like.
The intermediate code is a further reinscription of the source code, a
further 'exteriorization' of the software system, and a further transformation of the ruptures inscribed in the source program—ruptures that actually make it possible for code to function.

It is important to notice at this point that, while performing all
these reinscriptions of the source program, a compiler also checks
the program for correctness—that is, it assesses the program's com-
pliance with the rules of transformation defined for the specific programming language in which it was written.[47] However, a compiler
can only detect errors according to such rules (for instance, it can
easily locate a missing ';') but it cannot anticipate whether the
source program will function in an unpredictable way when executed. For instance, it cannot anticipate whether a perfectly correct

portion of the source program will cause the computer to enter an infinite loop with unforeseeable consequences. Let me take a little detour here in order to provide a further example. One of the most famous bugs in software—so famous that in the late 1980s professionals started wondering whether it was in fact just an urban myth—is the following FORTRAN statement:

DO 10 I = 1.10

...

10 CONTINUE

This legendary piece of code is largely quoted by technical literature as well as by the many websites devoted to the discussion of software errors. When executed, it leads to an infinite loop, which erroneously has been held responsible for the failed launch of Mariner I, the first American space probe, in 1962. In a posting on 'Famous Fortran Errors' dated 28 February 1995, John A. Turner debunks the myth of the FORTRAN error by referring to a previous discussion initiated by Dan Pop on Monday, 19 September 1994.[48] To give but one example of the many pieces of narrative circulating on the Internet in relation to this bug, Turner mentions how one of the Frequently Asked Questions on alt.folklore.computers used to be 'III.1—I heard that one of the NASA space probes went off course and had to be destroyed because of a typo in a FORTRAN DO loop. Is there any truth to this rumor?'[49] Also, Dieter Britz is reported to have posted the following question on the same newsgroup: '[i]t is said that this error was made in a space program, and led to a rocket crash. Is this factual, or is this an urban myth in the computer world?' Other incorrect (but often sharp and colorful) versions circulate, such as those posted by craig@umcp-cs ('[t]he most famous bug I've ever heard of was in the program which calculated the orbit for an early Mariner flight to Venus. Someone changed a + to a - in a Fortran program, and the spacecraft went so wildly off course that it had to be destroyed') and by Sibert@MIT-MULTICS ('it is said that the first Mariner space probe, Mariner 1, ended up in the Atlantic instead of around Venus because someone omitted a comma in a guidance program'), both reported by Dave Curry in his online piece on 'Famous Bugs'.[50] The legend results from the confusion of two separate events: Mariner I was in fact destroyed on 22 July 1962, when it started behaving erratically four minutes after launch, but such an erratic behaviour was due to a combination of hardware and software failures, while the DO-loop piece of code was identified (and corrected) at NASA in the summer of 1963 during the testing of a computer system.[51] In fact, according to the story originally recounted by Fred Webb in 1990 and subsequently reported by Turner, during the summer of 1963 a team of which Webb was part undertook preliminary work on the Mission Control Center computer system and programs. Among other tests, an orbit computation program that had previously been used during the Mercury flights was checked for accuracy, and the person conducting the tests came across a statement in the form of 'DO 10 I=1.10'.[52] Webb writes:

> This statement was interpreted by the compiler (correctly) as:
>
> DO10I = 1.10
>
> The programmer had clearly intended:
>
> DO 10 I = 1, 10
> After changing the `.' to a `,' the program results were corrected to the desired accuracy.[53]

Apparently, the program's answers had been accurate enough for the suborbital Mercury flights, so no one suspected a bug until the programmers tried to make it more precise in anticipation of later orbital and moon flights. Thus, the error seems to have been found under harmless circumstances and was never the cause of any actual failure of a space flight.

What I want to emphasize by retelling this story is how, if we follow the popular explanation, the misguided introduction of a dot ('.') rather than a comma (',') into the source program—a human error on the part of the programmer—resulted in a perfectly 'correct' program statement which could be 'tokenized' by the compiler according to the rules of substitution of FORTRAN language.[54] However, the dot functions as a rupture in the FORTRAN string and it leads to a very different tokenization than the one that would be obtained if the dot was replaced by a comma. With a comma, the string would be broken down by the compiler into the following tokens:

DO
10
I
=
1
10

With a dot, the string is in turn broken down into the following tokens:

DO10I

=

1.10

Both sequences of tokens are correct according to the rules of substitution of FORTRAN. However, when executed, the first sequence of token leads to the repetition of the loop for ten times, while the second leads to an infinite loop — that is, to the repetition of the loop which goes on *for ever*.[55] Thus, ultimately, a perfectly linearized string results in the perfect execution of a sequence of actions that potentially leads the computer system to disaster. In other words, execution (or, in the technical jargon, 'run time') is the moment in which apparently successful linearization leads to uncontrollable consequences.[56] Linearization does not therefore ensure the perfect calculability of the consequences of technology (and, broadly speaking, of the future). In order to explain how this happens, let me now return to my analysis of the compiler and explore how the source program is finally reinscribed as circuitry.

Eventually, the intermediate code generated by the compiler is reinscribed by the 'code generator' (the last phase of compiling) into 'the target language'. Aho et al. explain:

> The code generator takes as input an intermediate representation of the source program and maps it into the target language. If the target language is machine code, registers or memory locations are selected for each of the variables used by the program. Then, the intermediate instructions are translated into sequences of machine instructions that perform the same task.[57]

Importantly, this complex passage clarifies how, after all the above reinscriptions, the source code is now inscribed ('mapping' is just another term for 'reinscription', or 'replacement') in the computer memory in a form called 'machine code'. One must be reminded at this point that all the previous transcriptions were equally inscribed in the computer memory — for instance, in the initial string of the source code, a blank was encoded as the ASCII code for blank. But all these previous inscriptions were so complex that it would be impossible to follow them in this chapter. Now that we have reached a transcription in 'machine code', though, it is much easier to show how this string is inscribed and what happens during the subsequent stages of reinscription (the so-called 'execution', or 'run-time')—in other words, to find out what kind of reinscription execution is. However, the above passage also contains a number of cryptic terms such as 'registers', 'memory locations', and 'variables' that need to be explained before I can start discussing the reinscription of the program in machine code. Such an explanation, in turn, requires an understanding of the way in which what is usually called 'hardware' works. Let me then introduce a brief digression in order to clarify this point.

We have seen how the distinction between hardware and software is a shifting and unstable one. In Chapter 3, I showed how such a separation—just like the separation between different phases of software development—was also motivated by a division of labour. In the late 1960s, the distinction between hardware and software, as well as the conceptualization of hardware according to the so-called Von Neumann model, was almost universally accepted.[58] For the sake of simplicity, in the further section of this chapter I am going to utilize the Von Neumann's architecture, although it must be kept in mind that modern machines, while still adhering to this basic model, are significantly more complex. In this model, a computer is understood as composed of many different interconnected parts—namely, the input/output peripherals, the primary and secondary memory, and the Central Processing Unit (CPU).[59] The peripherals are devices such as keyboards, screens, and printers connected to and controlled by the computer but external to the CPU that allow for the transfer of information between the computer and the outside world. The CPU and the memory enable the storage and elaboration of data and programs. These parts are of particular importance to understand the workings of machine code. The CPU (or processor) basically consists of processing and control circuitry, and it is conventionally considered the computer's 'brain'.[60] Functionally, the CPU is divided into a control unit, which obtains instructions and sends signals to execute them, and the Arithmetic-Logic Unit (ALU), which executes mathematical computations and logic functions.

However, according to Von Neumann, without memory the CPU would be useless, since it only executes the operations described by a program, such as the one we have seen above, stored in the memory. The computer's primary memory is made up of a so-called RAM (Random Access Memory) and a ROM (Read Only Memory). The latter is programmed by the computer manufacturer and cannot be modified by computer users, including programmers, while the former is the primary working memory of the computer, where programmes and data are stored in a way that makes them easily accessible by the CPU. The RAM is a volatile memory: when the computer is switched off, all the data and programs residing in the RAM are lost. For this reason, there must exist a few permanent programs in the ROM that are executed when the computer is switched on again, in order to start up the whole system ('bootstrap programs').[61] The RAM and the ROM are both structured in 'cells', or 'memory locations', univocally identified by an address—namely, a progressive number usually expressed in binary or hexadecimal code. Every memory location 'contains' a value, expressed as a sequence of bits—but again the expression 'contains' is a metaphor. In fact, a memory location is not a box: it is a small logic circuit (called a 'flip-flop') which is designed in such a way that it preserves its value (either '0' or '1', the presence or absence of electrical current) over time.[62] The secondary memory comprises all those devices for the storage of information that are external to the primary memory (such as hard and floppy disks, CDs, and DVDs). Importantly, the ALU is also provided with small internal memories called 'registers', devoted to temporarily storing values and external memory addresses. The RAM and the internal memory of the CPU—that is, different types of registers—are essential to the understanding of code execution. Furthermore, all these hardware components are connected to one another through internal electrical pathways, called 'buses', along which binary signals are sent. In other words, a bus either transports ones (passage of current, also codified as TRUE) or zeroes (absence of current, also codified as FALSE). There are fundamentally three types of bus: the 'address bus' (which uni-directionally sends addresses from the CPU to the memory or to the peripherals), the 'data bus' (which sends data back and forth between the CPU and the memory or the peripherals) and the 'control bus' (which carries the control units' signals). In our simplified example, these three types of buses, the CPU registers and the logical circuitry of the ALU are all that is needed to explain whether and in what way the execution of code can be interpreted as a process of reinscription.

Now that I have illustrated the indispensable terms describing the computer circuitry, let me return to the example of a source program written in C language, in order to look at its reinscription in machine code. It is worth noting at this point that, after having been broken into lexical units, transformed into tokens and re-inscribed as intermediate code, the statement 'x=x+y;' that constitutes the central part of the programme

```
int x; int y; int z;
{
x=x+y;
}
```

has finally been reinscribed in the computer's secondary memory (for instance, on the hard disk) as something that, for the sake of clarity, we could represent with the following:[63]

1	MOV AX,[202]
2	ADD AX,[204]
3	MOV [200],AX
4	....
....	
200	10
202	3
204	4

As we have seen, the above language is called 'assembler'. Unlike binary code, assembler contains 'mnemonics', or strings (abbreviations or words) that make it easier for a human reader to remember a complex instruction.[64] We shall see in a moment how, for instance, the mnemonic MOV stands for a complex process that ultimately 'moves' data to a memory location. Every line of the program above is called an 'instruction'. It is worth noting here that the brief C statement 'x=x+y;' has become a long sequence of assembler instructions. However, and once again, the above list of six assembler instructions is just one possible reinscription of 'x=x+y;'. Another possible reinscription is as binary code—that is, as a sequence of binary digits, in which each assembler instruction actually corresponds to one binary instruction. Yet another possible reinscription is the detailed narrative that I offer in the following paragraphs. Again, it is important to remember that this narrative is significantly simplified in notation terms, since it makes use of assembler instructions rather than binary instructions. I will come back to this later on in the chapter.

In order to be executed, the program, which is stored in the secondary memory of the computer (for instance in the hard disk), is 'loaded' (that is, reinscribed) from the hard disk into the primary memory (RAM), starting from a certain memory address. This initial address is also copied into a register of the CPU called Program Counter (PC), which keeps track of the execution process by storing the address of the next instruction to be executed. The instruction currently being executed is reinscribed into another register of the CPU, the Instruction Register (IR). The program is executed one instruction at a time, from the beginning to the end. I want to emphasize here how the assembly program, the binary code, and the process of execution are still dominated by the same characteristics of linearity and discreteness that I have already pointed out in my analysis of formal grammars and programming languages. The assembler program is still sequenced in time. Moreover, for every single instruction, a number of reinscriptions are carried out in a sequence, which is called the 'Fetch-Execute Cycle'.[65] A technical manual would describe these actions as follows: the processor *fetches* (or 'loads') the first instruction from memory into the IR, it recognizes it as either an arithmetical or logical operation, or as a 'move' instruction, and it executes it by activating the apposite circuit and then by inscribing the result in another memory location. Thus, the Fetch-Execute Cycle consists of successive reinscriptions of bits and strings of bits *in/as* the computer circuitry. Even more importantly, the description of the Fetch-Execute Cycle that I have just provided takes the form of a narrative. Such a narrative is itself one of the possible re-inscriptions of the C instruction:

x=x+y;

and of the assembler instructions:

1	MOV AX,[202]
2	ADD AX,[204]
3	MOV [200],AX
4	....
....	
200	10
202	3
204	4

Let me now examine this narrative in more detail, in order to help us understand in what way the linearized sequence of assembler instructions leads to changes in the relevant computer circuitry and in what way these changes are ordered in time.

In the above assembler code, a memory location in the RAM is associated with each of the variables *x*, *y*, and *z*; while each memory location contains a bit string. We shall see in a moment that these bit strings are differently sequenced (technically, 'coded'), depending on the logical value of the variables. For the sake of clarity, let me also presuppose that *x*, *y*, and *z* respectively have the numerical values of '10', '3', and '4' before the execution of the program. After the execution, y and z will remain unchanged, while x will have the value of '7' (namely, the sum of 3 + 4). Let me also presuppose that the variables will be stored in the memory addresses 200, 202, 204. In the example provided above, AX is the name of a register of the CPU.

When the Fetch-Execute Cycle begins, the PC contains the address of the first instruction to be executed ('1'). This value ('1') is copied (in other words, reinscribed as a sequence of binary impulses) into the address bus and transported to the primary memory. In turn, the instruction 'MOVE AX,[202]', stored (inscribed) as a string of bits in the memory location identified by '1' is transported back to the CPU—which means that this instruction is reinscribed as a sequence of binary impulses (again, the binary *inscription* of the instruction 'MOVE AX,[202]') that travel on the physical pathway of the bus towards the CPU. Here the circuits that constitute the IR change accordingly, *becoming* the reinscription of the binary code of the instruction 'MOVE AX,[202]' itself. This is why I have stated above that bits and strings are inscribed *in/as* computer circuitry.[66] The fetch phase ends with the increment of the PC, which now has the value '2'—that is, the value of the next memory location.

The binary code for 'MOVE AX,[202]' is now inscribed in/as the IR, and it describes the action 'copy the value contained in the memory location 202 (which corresponds to the value of variable y) into the register AX'. To execute this instruction, the address 202 (coded as a string of bits) is transported on the address bus; subsequently, the content of the corresponding memory location (the binary encoding for '3') is retrieved and transported onto the data bus to the CPU and re-inscribed in the register AX. Now '3' is inscribed in binary form in/as the register AX. Thus the execution (and the Fetch-Execute Cycle) of the first instruction terminates.

Next, the Fetch-Execute Cycle of the second instruction begins. The value contained in the PC ('2') is transported from the CPU to the primary memory on the address bus. The instruction 'ADD AX,[204]' is carried to the CPU on the data bus and re-inscribed into the IR. The PC is incremented to '3', and the *fetch* phase terminates. The instruction 'ADD AX,[204]'—'add the value in the registry AX to the value in the memory location 204, and store the result in the registry AX'—is executed. This means that this time the address 204 travels from CPU to memory, through the address bus, and that the content of the memory location identified by 204 (the value '4') is transported back from memory to CPU, through the data bus. The arithmetical circuitry of the ALU calculates the 'ADD' operation on the two values '3' and '4' and the result ('7') is stored in the AX register. This means that now the circuitry constituting the AX register, which encoded the value '3', has its polarities changed to 'become' the binary encoding (inscription) of the value '7'. The execution of the second instruction is thus completed and the *fetch* phase of the third one begins.

Again, the value of the PC ('3') is transported to the memory on the address bus; the corresponding instruction 'MOV [200],AX' is transported from memory to CPU and copied in the IR, and the PC is incremented to '4'. The 'MOV [200],AX' ('copy the value of the registry AX into the memory location identified by 200') is executed. This time it is necessary to transport the value '200' on the address bus and the value '7' stored in AX on the data bus at the same time. It becomes clear at this point that all the above activities need to be coordinated in time—and for this reason the CPU contains a 'clock', or an electronic circuit that generates evenly spaced impulses *to keep track of time*. Ultimately, then, this is the internal computer time of program execution. Eventually, however, values '7' and '200' reach the central memory, and the binary code for '7' is stored in the memory location identified by '200'.

Importantly, the above re-inscription of the piece of code

1	MOV AX,[202]
2	ADD AX,[204]
3	MOV [200],AX
4	....
....	
200	10
202	3
204	4

takes the form of a detailed narrative precisely because it needs to give an account of the sequence of changes taking place within the computer circuitry—a sequence ordered *in time* and regulated by the computer internal clock (that is, a mechanism that keeps track of time through generating physical 'traces' in the form of evenly spaced impulses). Once again, though, such a 'narrative' is just another possible re-inscription of the assembly code and of the C statement 'x=x+y;'.[67] An alternative to such a narrative would be to 'observe' the physical changes of the relevant computer circuitry—although this is the point where software becomes 'unobservable'. This does not amount to saying that the above narrative is the 'representation' of something unperceivable that nevertheless 'really' happens in the computer, or that such a narrative is—in Alexander Galloway's terms—an extended 'metaphor' for the 'real', 'physical' level of circuitry.[68] Instead, it is important to acknowledge that our access to software is constantly mediated (by the technical literature on programming languages and compilers, by the written form of source code, by other reinscriptions such as the ones I am offering in this chapter).[69] The whole of these mediating reinscriptions present software as an incessant process of self-differentiation, or *différance*. To clarify this point further, let me now examine another possible reinscription of the above assembler code.

As I pointed out earlier on, the narrative I have offered is in itself quite simplified, because it does not take into account that everything in the computer is ultimately inscribed as strings of bits—that is, as polarized (on/off, '0'/'1') circuitry. Moreover, while memory is conventionally subdivided into bytes (standard strings of eight bits), what is usually considered as the minimum memory unit is not the byte but the 'word'. Here the meaning of 'word' is different from the one deployed by Hopcroft and Ullman, and it is conventionally considered as expressing a hardware characteristic. In fact, the word has the same capacity (or length) of the data bus and of the CPU registers. It usually coincides with the number of bytes used to memorize an integer number. Therefore, to add another layer to the self-differentiation of software, the values of the integers we have seen above ('3', '4' and '7') are inscribed in the computer as strings of bits. Similarly, the instructions of the above examples are also inscribed in the computer as strings of bits. The length of the string depends on the processor. For instance (in an INTEL 8086 processor), the instructions of the above example are encoded respectively on the following numbers of bytes:

MOV AX,[202]	3 bytes
ADD AX,[204]	4 bytes
MOV [200],AX	3 bytes

Thus, these assembler instructions can also be re-inscribed in one of the following ways:

Memory addresses	Machine language	Assembler language
0CA0:0100	A10202	MOV AX,[202]
0CA0:0103	03060402	ADD AX,[204]
0CA0:0107	A30002	MOV [200],AX

The above table simply means that the content of the memory address '0CA0:0100', which can be inscribed as 'MOV AX,[202]', can also be inscribed as 'A10202'. But again one must keep in mind that these are only *some* of the possible re-inscriptions of the string 'MOV AX,[202]'. In fact, both '0CA0:0100' and 'A10202' appear here in the so-called hexadecimal form. The hexadecimal form is nothing but an alphabet composed of sixteen different digits (the numbers from '0' to '9' plus the letters 'a', 'b', 'c', 'd', 'e', 'f') that can rein- scribe the decimal numbers according to Table 5.1:

Table 5.1.

Hexadecimal Notation	Decimal Notation
0	0
1	1
2	2
3	3
4	4
5	5
6	6
7	7
8	8
9	9
A	10
B	11
C	12
D	13
E	14
F	15
10	16
11	17
....	....

Simple algebraic operations allow the translation (reinscription) of
the hexadecimal notation into the binary one (see Table 5.2):

Table 5.2.

Hexadecimal Notation	Binary Notation
0	0000
1	0001
2	0010
3	0011
4	0100
5	0101
6	0110
7	0111
8	1000
9	1001
A	1010
B	1011
C	1100
D	1101
E	1110
F	1111

Table 5.2 shows how the symbol '1' of the hexadecimal alphabet can
be reinscribed as the string '0001' of the binary alphabet, and so on.
Therefore, the program of our example can be reinscribed as in
Table 5.3:

Table 5.3.

Memory Address	Machine Language	Assembler Language
0000110010100000:0000000100000000	1010000100000010000000010	MOV AX,[202]
0000110010100000:0000000100000011	00000011000001100000001000000010	ADD AX,[204]
0000110010100000:0000000100000111	1010001100000000000000010	MOV [200],AX

To understand Table 5.3 one must be reminded that one byte can encode two hexadecimal digits—that is, that the binary string '0001' is actually reinscribed in the computer memory as '00000001'. What appears obvious from the above table is that the binary representation of our program—albeit very close to the functioning of software *as* circuitry—is quite long and difficult to handle. Nevertheless, this was the situation before mnemonics were introduced, the assembler language was developed, and high-level languages were designed. I have therefore come full circle to the concept of programming language, formal grammar, and compiler that I have examined in Hopcroft and Ullman's textbook. After all, as we have seen at the beginning of this chapter, less than fifty years ago no high-level programming language was available. For instance, in order to be able to develop FORTRAN, Backus first needed to develop a formal grammar, or a method that allowed him to describe (or rather, inscribe) language in some form, so that it could be externalized on paper in a way that could also be used as a foundation for the development of a compiler. In other words, he needed a form that could be reinscribed many times in order to become software, which would then be capable of compiling a FORTRAN program.

However, it is hopefully quite clear now how the narrative that I have offered earlier on in this chapter while describing the Fetch-Execute Cycle constitutes in itself a somewhat abridged but meaningful reinscription of code. In fact, such a narrative makes visible in what way the linearized string of code (which undergoes many reinscriptions during the phases of compilation and execution, which in turn are made possible by its interruptions, ruptures, or 'blanks') 'becomes'—or, even better, *differentiates itself into*—a sequence of changes in computer circuitry. It also makes visible how the ruptures in code work, or, how the iterability of code—that is, its very repeatability—constitutes the time of code execution. Importantly, such a narrative also shows how writing about code is always self-reflexive. In other words, writing about code is always also a *reinscription* of code. In this sense, *Software Theory* can be seen as much a part of software as technical manuals, specifications, C programs, and binary code.

If we think of software as a process of constant self-differentiation, it makes no sense to ask whether software is really 'language', or 'code' (and what kind of code), or even 'just hardware'. In fact, it makes no sense to ask what software 'really' is. Instead, a more productive approach would be to engage in a deconstructive analysis of any of the singular forms that software takes. Such a process of deconstruction would aim at making visible the process of indefinite differentiation through which software is constituted. Saying that there is no 'truth' of software is tantamount to saying that there is no 'origin' of software—that is, no privileged point of view from which software can be apprehended once and for all. This is what I mean when I suggest that software is 'aporetic'. However, it is always possible to analyze the way in which software has become what it is—for instance, it is possible to examine the way in which compilers were developed in order to deal with the unworkable complexity of code. Such a deconstructive analysis would not be a narrative of the history of compilers or a technical exposition of how they function. It would rather take such historical and functional narratives as its point of departure in order to discover how they constitute an—always aporetic, always singular—instance of software. Performing such a process of deconstruction would be a politically meaningful gesture, because it would show on what conceptual points of opacity software rests, how it has become what it is, and what possibilities there might be to reinscribe those narratives differently.

Of particular political significance is the fact that a deconstructive reading sheds light on the different ways in which software gives rise to unexpected consequences. As we have seen, to rework Stiegler's terms, software is mnemotechnics which transgresses its aim (and therefore it is not fully thinkable as mnemotechnics). Software always implies iteration, which in turn implies variation through repetition. Thus, software always entails unpredictable effects. But, as I have shown earlier on in this chapter, the 'unexpected' can only be unexpected for someone—that is, from the point of view of a 'who', which is so deeply involved in its coevolution with the 'what' that it cannot foresee some of the possible self-differentiations of this process of coevolution. The moment when we become aware of some unexpected consequences of software—that is, the moment when we are 'surprised' by software—is the moment when we form a 'point of view' on software that aporetically separates ourselves from it. However, the most important unexpected consequences of software are not those that can be identified as straightforward malfunctions or failures (although these are also important, and perhaps also perceived as the most terrifying), but those that make it difficult for us to tell whether software is malfunctioning or not. These unexpected consequences (the moments in which technology appears to be 'quasi-functioning', and we are forced to ask: 'is it broken?', 'does it work?') are Derridean points of opacity, and they 'reveal' the underlying assumptions of software—the ones we rely upon in order to make software intelligible. Once these assumptions are transgressed (and assumptions will always be transgressed at some point, because they are inherently unstable) we need to make a decision—which is not a technical decision—in order to reestablish some intelligibility of technology (for example, we need to classify something as a malfunction and correct it, or to make space for a different understanding of technology that includes new interactions and new behaviours). Thus, such decision is always also an intervention. Points of opacity are also political opportunities. Software always interrogates the framework through which we make it intelligible, and it invites us to take action. But our interventions on software are always also reinventions—that is, they are part of the process of rediscovery of our originary technicity as human beings.

NOTES

1. Jacques Derrida, *Of Grammatology* (Baltimore: The Johns Hopkins University Press, 1976), 86.

2. Peter Naur and Brian Randell, eds., *Software Engineering: Report on a Conference Sponsored by the NATO Science Committee, Garmisch, Germany, 7th to 11th October 1968* (Brussels: NATO Scientific Affairs Division, 1969); John N. Buxton and Brian Randell, eds., *Software Engineering Techniques: Report on a Conference Sponsored by the NATO Science Committee, Rome, Italy, 27th to 31st October 1969* (Birmingham: NATO Science Committee, 1970). See also Brian Randell, "Software Engineering in 1968," *Proceedings of the IEEE 4th International Conference on Software Engineering* (Munich, Germany, 1979), and Robert W. Sebesta, *Concepts of Programming Languages* (London and New York: Pearson and Addison-Wesley, 2008).

3. For the distinction between technics and mnemotechnics, see Bernard Stiegler, *Technics and Time, 3: Cinematic Time and the Question of Malaise* (Stanford, CA: Stanford University Press, 2011), 131. For the idea of organized inorganic matter, see Stiegler, *Technics and Time 1*, 49.

4. Alfred V. Aho, Monica S. Lam, Ravi Sethi, and Jeffrey D. Ullman, *Compilers. Principles, Techniques, and Tools* (London and New York: Pearson and Addison-Wesley, 2007).

5. Sebesta, *Concepts of Programming Languages*, 5. The question of the accuracy of calculation performed on a computer is substantially a physical problem. Scientific applications typically have to deal with very large or very small numbers, whose length easily exceeds the fixed memory space for numerical data. Therefore a microprocessor and a programming language devoted to scientific applications must be able to store numbers in a way that takes into account all significant digits—that is, a way in which the decimals are not fixed but 'floating'. The first language ever developed for scientific applications was FORTRAN.

6. Randell, "Software Engineering in 1968," 3. Many participants in the NATO conferences were also involved in the design of new programming languages. One of the main technical developments at the time was in fact the design of ALGOL (with the watershed meeting in Munich held not long after the Garmisch conference) and its competition with FORTRAN (in which IBM had a vested interest that ultimately led to ALGOL's failure on the market) and Pl/1 (a 'universalistic' language whose development was the result of a certain entrenchment of interests between business and scientific communities) (Sebesta, *Concepts of Programming Languages*, 56–58).

7. John E. Hopcroft and Jeffrey D. Ullman, *Formal Languages and Their Relation to Automata* (Reading, MA: Addison-Wesley, 1969), v. An automaton is basically a formalism used to describe abstract systems and their transformations (cf. Aho et al., *Compilers*, 147). I will not enter into the detail of the theory of automata, since for the purposes of this chapter, I focus on the concepts of 'language' and 'grammar' in the theory of formal languages. Nevertheless, in the second section of this chapter, I will relate formal languages to automata and I will show how the theory of automata is especially important for compilers—that is, for programs that translate other programs to make their execution possible.

8. Noam Chomsky, "Three Models for the Description of Language," *IRE Transactions on Information Theory* 2, no. 3 (1956): 113–24.

9. In fact, the theory of programming languages is based on subclasses of Chomskyan grammars, that Chomsky names 'context-free grammars' and 'context-sensitive grammars'.

10. Hopcroft and Ullman, *Formal Languages*, 1.

11. Bernard Stiegler, *Technics and Time, 1: The Fault of Epimetheus* (Stanford, CA: Stanford University Press, 1998), 3.

12. Hopcroft and Ullman, *Formal Languages*, 1.

13. Hopcroft and Ullman, *Formal Languages*, 1.

14. Cf. Derrida, *Of Grammatology*, 85–86.

15. Hopcroft and Ullman, *Formal Languages*, 1.

16. See also Aho et al., *Compilers*, 118.

17. Katherine N. Hayles, *My Mother Was a Computer: Digital Subjects and Literary Texts* (Chicago: University of Chicago Press, 2005). I have discussed Hayles's argument on the differences between natural and programming languages in Chapter 2.

18. Hopcroft and Ullman, *Formal Languages*, 1.

19. Hopcroft and Ullman, *Formal Languages*, 1.

20. Hopcroft and Ullman, *Formal Languages*, 1.

21. Hopcroft and Ullman, *Formal Languages*, 1–2.

22. This fact is inscribed in the name of the computer program Yacc, first developed in 1970. Yacc is mainly a parsing tool (whose functioning I shall discuss below) and its name stands for 'Yet Another Compiler Compiler'.

23. Chomsky, "Three Models for the Description of Language."

24. Hopcroft and Ullman, *Formal Languages*, 9.

25. Hopcroft and Ullman, *Formal Languages*, 9.

26. Hopcroft and Ullman, *Formal Languages*, 10.

27. Aho et al., *Compilers*, 199.

28. Hopcroft and Ullman, *Formal Languages*, 10.

29. Aho et al., *Compilers*, 198.

30. Aho et al., *Compilers*, 198.

31. Hopcroft and Ullman, *Formal Languages*, 18–19.

32. Jay David Bolter, *Turing's Man: Western Culture in the Computer Age* (London: Duckworth, 1984).

33. Cf. Jacques Derrida, *Limited Inc.* (Evanston: Northwestern University Press, 1988).

34. Hopcroft and Ullman, *Formal Languages*, 19.

35. Hopcroft and Ullman, *Formal Languages*, 20.

36. Interestingly, various parsing algorithms used in compilers have names like "LL parser", "LR parser", and "LALR parser", which make reference to left, right and 'ahead'. Thus, parsing algorithms are distinguished exactly by the nature of the way they traverse a tree.

37. The BNF is a widely used notation for the description of grammars—and specifically for the so-called context-free grammars. It was presented by John Backus at the first World Computer Congress in Paris in 1959 as a formal description of what would later become the ALGOL programming language. Later on Peter Naur proposed the name 'Backus Normal Form' for such notation, and he worked on it in order to simplify it. His name was then added to the name of the grammar in recognition of his contribution—the Backus Normal Form thus becoming known as the Backus Naur Form.

38. Friedrich A. Kittler, "There Is No Software," *CTheory* (1995), http://www.ctheory.net/articles.aspx?id=74.

39. Aho et al., *Compilers*, 1.

40. Aho et al., *Compilers*, 1.

41. This is again somehow an oversimplification, since not all compilers translate source code into binary code. For instance, Java compilers translate source code into Java bytecode, which can then be interpreted by a Java Virtual Machine.

42. Aho et al., *Compilers*, 5–6.

43. What counts as a rupture varies significantly from language to language. For Python, for example, white spaces at the beginning of a line are semantically significant and thus they do not function as ruptures.

44. Tokenization is in fact the first step of 'meaning-making', since it names a class of strings (variable name, arithmetic value, parenthesis, etc.) to which the 'original text' is attached.

45. Aho et al., *Compilers*, 8.

46. Aho et al., *Compilers*, 9.

47. Aho et al., *Compilers*, 1.

48. John A. Turner, "Famours Fortran Errors," http://www.rchrd.com/Misc-Texts/Famous_Fortran_Errors .

49. See http://faqs.cs.uu.nl/na-bng/alt.folklore.computers.html.

50. David Curry, "Famous Bugs," http://www.textfiles.com/100/famous.bug. Dave Curry's posting gives a detailed account of the research originally started by John Shore on 'documented reports on "famous bugs"'.

51. Paul Ceruzzi, *Beyond the Limits: Flight Enters the Computer Age* (Cambridge, MA: MIT Press, 1989), 202–3.

52. Project Mercury's suborbital flights took place in 1961 and its orbital flights began in 1962.

53. Fred Webb, "Famous Fortran Errors," http://www.rchrd.com/Misc-Texts/Famous_Fortran_Errors.

54. This narrative implies a slightly different functioning of FORTRAN lexical analyzers than the one I have described above. However, and notwithstanding its problematic aspects, such narrative has become a widely accepted explanation for this bug.

55. Technically, the FORTRAN statement containing the comma means 're-peat the action (represented by "... CONTINUE") while increasing the value of the counter "I" from 1 to 10, then stop', while the statement containing the dot means 'assign the value 10.1 to the variable "DO10I", then go on repeating the action "... CONTINUE" forever.' However, there is no need to recur to an explanation based on meaning to clarify the different executions of the two different pieces of code. As I will show in the rest of this chapter, the tokeniza-tion leads to different reinscriptions of the FORTRAN statement—that is, to different configurations of the circuitry of the system resulting in different com-puter behaviours.

56. The circular time of the loop is still linear time. This clarifies further how linearization cannot completely rule out the unexpected.

57. Aho et al., *Compilers*, 10.

58. The Hungarian-born mathematician John Von Neumann produced his key report on the structure of computers in 1945.

59. Mark Burrell, *Fundamentals of Computer Architecture* (Basingstoke and New York: Palgrave Macmillan, 2004), 5.

60. A processor constructed entirely as a very large electrical circuit (called an *integrated circuit*) on one single chip of silicon (colloquially called a *chip*) is called a *microprocessor*. What we call a 'computer' these days would be more accurately named as a *microprocessor-based computer system*, or micro-computer (Cf. Burrell, *Fundamentals of Computer Architecture*, 5).

61. Interestingly, the term 'bootstrap' refers to the famous literary episode in which the Baron of Munchausen lifted himself in the air by pulling at his own boots' straps.

62. Burrell, *Fundamentals of Computer Architecture*, 110.

63. Importantly, this is just one possible assembler notation. Different hard-ware has different associated assembly languages.

64. This is in fact how the assembler language was created: it was a way to substitute (or reinscribe) complex strings of binary digits with the more human-friendly strings called mnemonics.

65. Burrell, *Fundamentals of Computer Architecture*, 135.

66. Once again, it must be noticed that we have not reached a conclusive point in the narrative. As we have seen, the changes involved in the circuits of the IR are flip-flop switches from 0s to 1s and vice versa. However, it would be possible to analyze deeper changes in structure to help explaining the reinscrip-tion of machine code into sequences of circuit-level operations. The level of reinscription we are dealing with at this stage ('microcode') is substantially another way of abstracting machine instructions from the underlying electron-ics. Importantly, this again reflects a division of labour, since the design of microcode (sometimes called microprogramming) for a particular processor im-plementation is often done by an engineer at the stage of processor design. Microcode is generally not visible to a 'normal' programmer, not even to pro-grammers who deal with assembler code. It is also strictly associated with the particular electronic circuitry for which it is designed—in fact, it is an inherent part of that circuitry. In 1967, Aschler Opler introduced the term 'firmware' to indicate the microcode involved in the implementation of machine instructions (Aschler Opler, "Fourth-Generation Software," *Datamation* 13, no. 1 [1967]: 23). The term 'firmware' obviously denotes something that is neither hard nor soft, and it is a poignant indicator of the difficulty encountered by engineers when trying to distinguish different 'levels of materiality' in computers.

67. Although with the term 'narrative' I refer to my description of the Fetch-Execute Cycle, the assembly code could also be thought of as a piece of narra-

tive. My description of the Fetch-Execute Cycle—that is, the story I tell about code—reinscribes in a natural language (English) the sequencing of time that in the assembly code (or even in the C code) is inscribed as a string.

68. Alexander Galloway, *Protocol: How Control Exists after Decentralization* (Cambridge, MA: MIT Press, 2004).

69. The narrative of the CPU's functioning that I have just provided makes apparent how the interruptions in the string of the source program are reinscribed as changes of voltage ordered in time. Nevertheless, it would not be correct to assume that only such a narrative (or the 'actual' execution of code in/ as computer circuitry) constitutes a 'proper' account of software, and even less that only such a narrative accounts for the way in which the 'blanks' lead to the tokenization of code and, ultimately, to its execution at run-time. Rather, temporality is always present in the reinscriptions of code—for instance, in the orientated arrow of the transformation rules of formal grammars, or in the space of the page which presupposes the orientated movement of the eyes of the reader.

Conclusions

The Unforeseen Consequences of Technology

'Fighting an alien robot? That was me! And it was amazing!' boasts Susan Murphy soon after having defeated an alien robot probe in San Francisco with the help of a gelatinous blue blob and a gay fish-ape hybrid. She gleefully proceeds to enumerate all the different ways in which being a monster is an extremely appealing and re-warding status for an American girl of her age. All the while, the group of freaks that surround her—which includes a mad scientist and a perambulating insect chrysalis—marvel at the discovery of their own virtues and talent.

I would like to propose a brief reading of the computer-animat-ed feature film from DreamWorks, *Monsters vs. Aliens*, as a kind of 'alternative summary' of *Software Theory* here. *Monsters vs. Aliens* was released in March 2009. In this film, Susan Murphy, a young woman from Modesto, California, is hit by a radioactive meteor on the day of her wedding, thus absorbing a rare substance called quantonium that mutates her into a giantess. Immediately captured by the US military and classified as a 'monster', she is imprisoned in a top-secret facility directed by General W. R. Monger where other 'monsters' are kept in custody, among whom are B.O.B. (Bicarbo-nate Ostylezene Benzonate, an indestructible gelatinous blue blob without a brain), Dr. Cockroach, PhD (a mad scientist with a giant cockroach's head), the Missing Link (a twenty-thousand-year-old amphibious fish-ape hybrid), and Insectosaurus (a 350-foot grub). When an alien named Gallaxhar attacks the Earth with his gigantic robotic probes and an army of clones of himself, General Monger persuades the president of the United States to deploy the monsters as military weapons. Having accepted the mission with the promise of freedom if they succeed, the monsters manage to destroy the alien robotic probe that Gallaxhar has sent to San Francisco. During the fight Susan discovers that she possesses an unexpected strength and that she is also invulnerable to Gallaxhar's weapons. Having been freed, Susan happily returns to Modesto—only to be rejected by her fiancée (who claims that he cannot be married to a woman who overshadows him) while, unwittingly, her monstrous friends disseminate panic in the neighbourhood. Initially sad and dispirited, Susan suddenly realizes that becoming a monster has actually enriched her life, and she fully embraces her new 'amazing' lifestyle and her newly formed bond with the other monsters. After a final epic fight Susan and her gang completely defeat Gallaxhar and his cloned army, and are eventually acclaimed as heroes. In the last scene of the film, they are alerted to the fact that in the surroundings of Paris a snail has fallen into a nuclear power plant and is growing into a giant due to nuclear irradiation. They then fly off on a mission to protect the Earth from the new enemy.

In the context of the overall argument of my book, what is particularly interesting about *Monsters vs Aliens* is that in the movie the monsters function first and foremost as a figure of the unexpected consequences of technology.[1] In fact, they all *come into existence* as unforeseen effects of technology: they are the unpredictable outcomes of experiments gone wrong. B.O.B. was mistakenly created by injecting a genetically modified tomato with a chemically altered ranch dressing. Dr. Cockroach ended up with an insect head and the ability to climb walls while subjecting himself to an experiment in order to gain the longevity of a cockroach, and the mad scientist is the figure of the experiment gone wrong *par excellence*. Insectosaurus, originally a one-inch grub, was transformed into a giant after being accidentally invested by nuclear radiation. Even the Missing Link could not have been found frozen in a lagoon and thawed out by scientists without some help from technology.[2]

However, in *Monsters vs Aliens* the monsters are also 'domesticated'—or rather, they are kept under custody by the American government and later on transformed into weapons. In other words, the film seems to imply that technology needs to be controlled in order to be made useful—that is, it has to become a tool. But in order to be successfully deployed as weapons, monsters must be released from custody—that is, in order to be 'used', technology must be set free. In turn, once set free, technology escapes instrumentality. In fact, it is by fighting Gallaxhar that Susan discovers her unexpected physical strength. Similarly, during the final battle against the aliens Insectosaurus apparently dies, only to undergo a metamorphosis from a chrysalis into a beautiful butterfly. Ultimately, though, the monsters are still kept under control: they constitute an American military team, albeit a very special one. It is here that aliens find their place in the film narrative: a relationship which would otherwise be quite uncomplicated (humans detain and domesticate dangerous monsters) finds its third term in the aggressive threat from the outside. Aliens provide an enemy and help construct the narrative of the American fight for democracy against (alien) totalitarian regimes. Even though it occasionally makes fun of the American government (General W. R. Monger's name is a pun on the word 'warmonger', while the inept president of the United States is always on the verge of launching a nuclear attack by pressing the wrong button), the film still embraces a narrative that legitimates the Unites States as the world superpower.

The point of opacity of the film is to be found at its end: in the final scene the monsters set off to Paris to fight the gigantic snail which has broken into a nuclear plant—but should the snail be understood as an alien or a monster? Since it is presented as a threat against which the monsters are supposed to fight, it must be an alien. And yet, since clearly it is an unexpected effect of technology (actually, accidental nuclear irradiation is one of the most common origin stories of superheroes and is very similar to Insectosaurus's story), the snail must be a monster and in principle it should not be fought but rather helped out or maybe even recruited as part of the team. With a revealing lapse, the Wikipedia entry on *Monsters vs. Aliens* recounts how at the end of the film 'the monsters are alerted to a *monster attack* near Paris and fly off to combat the new menace' (emphasis added).[3] In Derridean terms, it could be said that the snail is the incest taboo of *Monsters vs Alien*.[4] In other words, the snail is the point where the instrumentality of technology undoes itself, because technology is always both a monster and an alien, an instrument and a threat, a risk and a promise. The unexpected is always implicit in technology, and the potential of technology for generating the unexpected needs to be unleashed in order for technology to function *as* technology. The attempt to control the unexpected consequences of technology is ultimately destined to fail— and yet it must be pursued for technology to exist. For this reason, every choice we make with regard to technology always implies an assumption of responsibility for the unforeseeable. As a deconstructive approach to software like the one I have offered in *Software Theory* shows, whenever one makes decisions about technology, one has to remember that technology can always generate consequences that escape predictability. Thus, to think of technology in a political sense we must first and foremost remember that technology cannot be thought from within the conceptual framework of calculability and instrumentality.

In *Software Theory*, I have proposed a radical rethinking of software as always entangled with the conceptual framework of instrumentality—typical of Western philosophy's understanding of technology—and thus somewhat complicitous with it, while at the same time capable of escaping it by transgressing all the conceptual boundaries associated with instrumentality (such as hardware/software, source code/machine code, program/process, language/materiality, technology/society). For instance, I have showed how the discipline of software engineering emerged as a strategy for the industrialization of the production of software at the end of the 1960s and how it understood software as a process of material inscription that continuously opened up and reaffirmed the boundaries between 'software', 'writing', and 'code'. Software engineering established itself as a discipline precisely through an attempt to control the constitutive fallibility of software-based technology. I have also presented the emergence of the free/open source movement in the 1990s as one of the unforeseen consequences of the software engineering of the 1970s and 1980s. Ultimately, I have argued that software engineering is characterized by the unstable definition of the instrumentality of software. In software engineering, the instrumentality of software is—as Derrida would have it—'in deconstruction': it is the unstable result of the process of technological exteriorization. And yet, since the undoing and redoing of instrumentality can go unnoticed, a deconstructive reading—that is, a critical and creative problematization—of software must be actively performed. This active problematization of software ultimately clarifies the significance of software—always to be thought in its singularity—for our understanding of the human and of its constitutive relationship with technology.

In the last section of *Software Theory*, I have taken a deconstructive approach to the theory of formal languages in order to shed light on what I have proposed to call the 'aporetic' nature of software. I have argued that software emerges as a self-differentiating process of material inscription and as a precarious process of linearization which continually transgresses the very conceptual categories through which it comes into existence. Open to iteration, which implies variation through repetition, software always entails unpredictable consequences—that is, unforeseeable aporetic differentiations of what Bernard Stiegler has named the 'who/what complex'.

Every time software brings about some unexpected consequences, it is fundamental to decide whether this is a malfunction that needs to be fixed or an acceptable variation that can be integrated into the technological system, or even an unforeseen anomaly that will radically change the technological system for ever. This is the fundamental double valence of the unexpected as both failure and hope. Like Derrida's *pharmakon*, technology entails poison and remedy, risk and opportunity.[5] Once again, the inseparability of these aspects means that, every time we make decisions about technology, we are taking responsibility for uncalculable risks. The only way to make politically informed decisions about technology is not to obscure such uncalculability.

In the documentary *The Net* (2003), director Lutz Dammbeck shows how obscuring the incalculability of technology leads to setting up an opposition between risk and control and between 'good' and 'bad' technology, and ultimately to the authoritarian resolution of every dilemma regarding technology. Questions such as 'Should technology be 'democratized'?', 'Should it be made available to everyone even when it is "dangerous"?', 'Who decides what is dangerous for whom?' are then addressed by embracing either a policy of control or a deterministic, almost paranoid fear of technology, which is also possibly combined with a Luddite stance. The film explores the complex story of Ted Kaczynski, the infamous Unabomber. A former mathematician at Harvard, Kaczynski retreated to a cabin in the wilderness of Montana in 1971. In 1996, he was arrested by the FBI under the suspicion of being responsible for the attacks carried out between 1978 and 1995 by an unknown individual nicknamed the Unabomber against major airlines executives and scientists at elite universities. The film complicates the narrative regarding the Unabomber (the author of an anti-technology *Manifesto*, and an ultimate figure of resistance for those who oppose contemporary technology as a form of control) by situating him within the complex and contradictory web of late twentieth-century technology.

Particularly revealing is an interview with John Taylor—an ex-NASA engineer and an admirer of Norbert Wiener, the founding father of cybernetics—which shows how the idea of calculability (and the attempt to expel the unexpected from technology) was crucial for early cybernetics. Taylor recounts how ARPA (the Advanced Research Projects Agency) was set up in 1958 by the American president Eisenhower with the goal of seeking out 'promising' research projects—in Taylor's words, projects that had 'a longer term expectation associated with them'. ARPA was instituted after the launch of the Russian space probe *Sputnik* in 1957, which Taylor characterizes as 'a great surprise' for the United

States. The US Department of Defense set up ARPA 'in the hope that we would not get surprised again like the Russians surprised us'. The ambivalence of the term 'surprise' as both risk and promise is obvious in Taylor's words: the best research projects are the ones which hold the 'promise' of 'good surprises', which will in turn prevent the enemy from surprising us in a 'bad' way. ARPA was therefore meant to 'domesticate' the potential of technology to surprise us, that is its capacity for generating the unexpected, by subjecting 'promising' projects to control. Taylor ostensibly embraces such a philosophy of control. When, during the interview, Dammbeck mentions the Unabomber, a horrified look crosses Taylor's face and, as many of his colleagues interviewed in the film do, he refuses to speak about Kaczynski, dismissing him as a terrorist and even comparing the Unabomber's *Manifesto* to Hitler's *Mein Kampf*. When Dammbeck suggests that some people such as the Unabomber might be scared by technology and asks Taylor what he is scared of, he answers 'I am scared of Al-Qaeda . . . I am scared of cancer. But if we could find a cure for cancer, we wouldn't be afraid.' According to Taylor, fear is a matter of ignorance, of 'not knowing'. By possessing more knowledge—he adds with a curious phrasing—we could 'prohibit cancer'. Taylor's revealing formulation is the ultimate expression of a desire for the technological control over nature and for the complete calculability of the future.

The idea of cybernetics as the science of control takes up a new meaning here—one related to prediction, calculation, foreseeability—if one considers, as Dammbeck does, that one of the participants in the Macy Conferences (which instituted cybernetics as a discipline between 1946 and 1953), the psychologist Kurt Lewin, conceived a project for programming humans to give them an 'anti-authoritarian personality', thus obstructing the possibility of fascism forever. Oblivious to the fact that this would be the ultimate authoritarian gesture, Lewin suggested that cybernetics could control and remap people's subconscious in order to immunize people against totalitarianism and to make authoritarian systems impossible. For him, anti-authoritarianism was first and foremost a matter of calculation, as the *control of the political future* of humanity.[6] Ironically, drawing on Lewin's project, Henry A. Murray, one of the fathers of today's assessment centres, devised a series of tests which were supposed to highlight concealed psychological tendencies by penetrating consciousness with non-surgical means—basically LSD and other drugs. Such tests were carried out by the CIA in the late 1960s at Harvard on a group of talented young male students, among whom was Ted Kaczynski. Whether those experiments led Kaczynski to the fear of occult forms of mind control, and ultimately resulted in his paranoid terror of technology is a possibility that the film leaves open. Importantly, however, Dammbeck's film makes a suggestion that control and uncalculability, risk and opportunity, are constitutive of technology. As Dammbeck himself states, the key to Kaczynski's tragedy is the fact that he is 'part of a system from which there is no escape'. He does not understand that, even isolated in a forest cabin, one is still part of the technological system (a cabin *is* a form of technology, after all), and that there is no 'outside' of technology.

Once again I want to emphasize here that in order to make responsible decisions about technology, one must be aware that technology (as well as the conceptual system on which it is based) can only be problematized from within.[7] In fact, the problem I started from, or the question that, following Stiegler, I emphasized in the Introduction to *Software Theory*—namely, that we need to make decisions about a technology which is always somehow opaque—requires precisely such an active problematization of technology. One must acknowledge that software is *always* both conceptualized according to a metaphysical framework *and* capable of escaping it, that it is instrumental *and* generative of unforeseen consequences, that it is both a risk and an opportunity. Such a problematization of technology is a creative, productive, and politically meaningful process. By opening new possibilities and foreclosing others, our decisions about technology also affect our future. Thus, thinking politics *with* technology becomes part of the process of the reinvention of the political in our technicized and globalized world.

NOTES

1. Judith Halberstam has proposed a 'queer' archive of animated features where the term 'queer' means that such features incorporate a politically subversive narrative cleverly disguised in a popular media form aimed at children (cf. Judith Halberstam, *The Queer Art of Failure* [Durham and London: Duke University Press, 2011]). For instance, according to Halberstam, the CGI animated film of 2003, *Finding Nemo*, depicts the title character—a motherless fish with a disabled fin—as a 'disabled hero' and links the struggle of the rejected individual to larger struggles of the dispossessed (Nemo leads a fish rebellion against the fishermen). Halberstam proposes the term 'Pixarvolt' to indicate movies depending upon Pixar technologies of animation and foregrounding the themes of revolution and transformation. For Halberstam, the Pixarvolt films use the individual character as a gateway to stories of collective action, anticapitalist critique, and alternative imaginings of community and responsibility. In a

sense, it could be said that the monsters in *Monsters vs Aliens* yield themselves to a queer reading—actually, queer references seem to have become quite commonplace in animated features. For instance, the Missing Link is a parody of excessive masculinity (notwithstanding his machismo and his gung-ho attitude to fight, he is comically out of shape) and has a gay bond with Insectosaurus; the monsters perform part of their first battle against Gallaxhar on a stolen San Francisco bus directed to the Castro; and, even more tellingly, the transformation of Susan into a monster frees her from all heterosexual social expectations and places her in a queer alliance within other social outcasts. Nevertheless, it is debatable whether the narrative of the film can be read as subversive, since the monsters' community seems not so much to constitute an alternative to the mainstream society as a weapon in the hands of the American government—although one could argue that such subversive narratives are at their most intriguing when they are apparently neutralized. As I will show in a moment, the neutralization of the monsters in *Monsters vs. Aliens* is in fact apparent, but to understand this point the monsters must be viewed in their relationship with technology.

2. Ostensibly the film here taps into the popular tradition of superheroes that has dominated American comic books for decades and has subsequently crossed over into other media. The so-called 'origin stories' associated with superheroes, which explain the circumstances by which the characters acquired their exceptional abilities, often involve experiments gone wrong. For instance, Spider Man (Peter Parker) got bitten by a radioactive spider during a science demonstration at school when he was a teenager, while the Fantastic Four (Reed Richards, Sue and Johnny Storm, and Ben Grimm) were accidentally exposed to cosmic rays during a space mission (Richard Reynolds, *Super Heroes: A Modern Mythology* [Jackson: University Press of Mississippi, 1994]). In other words, it can be said that superheroes themselves embody unexpected consequences of technology, which have nevertheless been reframed into a narrative of fight against evil.

3. "Monsters vs Aliens," http://en.wikipedia.org/wiki/Monsters_vs_Aliens.

4. In *Of Grammatology*, Derrida famously shows how the incest taboo is the unthought of structural anthropology—that is, a concept that cannot be thought within the conceptual system of the discipline because it escapes its basic opposition between nature and culture (cf. Jacques Derrida, *Of Grammatology* [Baltimore: The Johns Hopkins University Press, 1976]).

5. Jacques Derrida, *Dissemination* (Chicago: University of Chicago Press, 1981).

6. This is just one of the many examples of how the concept of the 'system' of cybernetics was transferred to the social and political realm quite uncritically, starting from the 1950s. In the BBC documentary *The Trap* (2007), director Adam Curtis has shown how the idea of freedom that characterizes today's neoliberal democracies is a very limited one, mainly founded on the idea of the individual as a free agent always pursuing its own self-interest. This idea is very much based on game theory and other theories developed during the Cold War period, which had a strategic importance in determining the so-called 'balance of terror' (in which basically the enemy avoids attacking for fear of being destroyed). Driven by the necessity of anticipating Soviet moves, John F. Nash's games were based on the belief that, *in every society*, stability could be created through distrust. Nash's equations work only if individuals are presumed to be selfish and suspicious of one another. If they start to cooperate, however, then the system becomes unpredictable. In the famous Prisoner's Dilemma, it is selfishness that leads to safety. This game shows that the rational move is always to

betray the other. This was the logic of the Cold War (that is, one cannot trust the other not to cheat)—but Nash turned this assumption into a *theory of society*. In his paranoid view of the lonely human being in a hostile society, the price of freedom is distrust. Curtis's fascinating documentary shows how Nash's ideas spread to fields that had nothing to do with nuclear strategy, from R. D. Laing's psychiatric thought to George Buchanan's theories of 'public choice' which assisted Margaret Thatcher's dismantling of the welfare state.

7. This is precisely what deconstruction allows us to do—stepping out of a conceptual system by continuing to use its concepts while at the same time demonstrating their limitations.

Bibliography

Aho, Alfred V., Monica S. Lam, Ravi Sethi, and Jeffrey D. Ullman. *Compilers. Principles, Techniques, and Tools*. London and New York: Pearson and Addison-Wesley, 2007.

Aho, Alfred V., and Jeffrey D. Ullman. *Principles of Compiler Design*. Reading, MA: Addison-Wesley, 1979.

Alt, Franz. L., and Morris Rubinoff, eds. *Advances in Computers*. New York: Academic Press, 1968.

Aristotle. *The Complete Works*. Princeton, NJ: Princeton University Press, 1968.

Austin, John L. *How to Do Things with Words*. Oxford: Clarendon Press, 1972.

Barad, Karen. "Posthumanist Performativity: Toward an Understanding of How Matter Comes to Matter." *Signs: Journal of Women in Culture and Society* 28, no. 3 (2003): 801–31.

Beardsworth, Richard. *Derrida and the Political*. New York: Routledge, 1996.

———. "From a Genealogy of Matter to a Politics of Memory: Stiegler's Thinking of Technics." *Tekhnema: Journal of Philosophy and Technology* 2 (1995): 85–115.

Belady, Laszlo A., and Meir M. Lehman. "A Model of Large Program Development." *IBM Systems Journal* 15, no. 3 (1976): 225–52.

Berry, David. *The Philosophy of Software: Code and Mediation in the Digital Age*. Basingstoke: Palgrave Macmillan, 2011.

Bolter, Jay David. *Turing's Man: Western Culture in the Computer Age*. London: Duckworth, 1984.

Bolter, Jay David, and Richard Grusin. *Remediation: Understanding New Media*. Cambridge, MA: MIT Press, 2002.

Braudel, Fernand. *Capitalism and Material Life 1400–1800*. London: Weidenfeld and Nicolson, 1973.

Brooks, Frederick. P. *The Mythical Man-Month: Essays on Software Engineering, Anniversary Edition*. Reading, MA: Addison-Wesley, 1995.

———. "No Silver Bullet: Essence and Accidents of Software Engineering." *IEEE Computer* 20, no. 4 (1987): 10–19.

Burrell, Mark. *Fundamentals of Computer Architecture*. Basingstoke and New York: Palgrave Macmillan, 2004.

Butler, Judith. *Excitable Speech: A Politics of the Performative*. New York and London: Routledge, 1997.

Buxton, John N., Peter Naur, and Brian Randell, eds. *Software Engineering: Concepts and Techniques*. New York: Petrocelli-Charter, 1976.

Buxton, John. N., and Brian Randell, eds. *Software Engineering Techniques: Report on a Conference Sponsored by the NATO Science Committee, Rome, Italy, 27th to 31st October 1969*. Birmingham: NATO Science Committee, 1970.

Campbell-Kelly, Martin. *From Airline Reservations to Sonic the Hedgehog: A History of the Software Industry*. Cambridge, MA and London: MIT Press, 2003.

Ceruzzi, Paul. *Beyond the Limits: Flight Enters the Computer Age*. Cambridge, MA: MIT Press, 1989.

———. *A History of Modern Computing*. Cambridge, MA and London: MIT Press, 2003.

Chomsky, Noam. *Aspects of the Theory of Syntax*. Cambridge, MA: MIT Press, 1965.

———. "On Certain Formal Properties of Grammars." *Information and Control* 2, no. 29 (1959): 137–67.

———. *Syntactic Structures*. The Hague: Mouton, 1957.

———. "Three Models for the Description of Language." *IRE Transactions on Information Theory* 2, no. 3 (1956): 113–24.

Chun, Wendy Hui Kyong. *Control and Freedom: Power and Paranoia in the Age of Fiber Optics*. Cambridge, MA and London: MIT Press, 2006.

———. *Programmed Visions: Software and Memory*. Cambridge, MA and London: MIT Press, 2011.

Clark, Timothy. "Deconstruction and Technology." In *Deconstructions. A User's Guide*, edited by Nicholas Royle, 238–57. Basingstoke: Palgrave, 2000.

Clocksin, William. F., and Christopher S. Mellish. *Programming in Prolog*. New York: Springer-Verlag, 2003.

Constantine, Larry L. "The Programming Profession, Programming Theory, and Programming Education." *Computers and Automation* 17, no. 2 (1968): 14–19.

Conway, Melvin. E. "How Do Committees Invent?" *Datamation* 14, no. 4 (1968): 28–31.

Cox, Brad J. "There Is a Silver Bullet." *BYTE Magazine*, October, 1990.

Cramer, Florian. *Anti-Media: Ephemera on Speculative Arts*. Rotterdam: NAi Publishers and Institute of Network Cultures, 2013.

Culler, Jonathan. *Ferdinand de Saussure*. New York: Cornell University Press, 1986.

Curry, David. "Famous Bugs." http://www.textfiles.com/100/famous.bug.

Dahl, Ole-Johan, Edsger W. Dijkstra, and Charles A. R. Hoare. *Structured Programming*. London: Academic Press, 1972

Derrida, Jacques. *Archive Fever: A Freudian Impression*. Chicago: University of Chicago Press, 1996.

———. *Dissemination*. Chicago: University of Chicago Press, 1981.

———. *Edmund Husserl's Origin of Geometry: An Introduction*. New York: Nicolas Hays, 1978.

———. "Letter to a Japanese Friend." In *Derrida and Différance*, edited by Robert Bernasconi and David Wood, 1–5. Warwick: Parousia Press, 1985.

———. *Margins of Philosophy*. Chicago: Chicago University Press, 1982.

———. *Mémoires: for Paul de Man*. New York: Columbia University Press, 1986.

———. *Of Grammatology*, Baltimore: The Johns Hopkins University Press, 1976.

———. *Paper Machine*. Stanford, CA: Stanford University Press, 2005.

———. *Points. . . : Interviews, 1974–1994*. Stanford, CA: Stanford University Press, 1995.

———. *Positions*. London and New York: Continuum, 2004.

———. *The Post Card: From Socrates to Freud and Beyond*. Chicago: University of Chicago Press, 1987.

———. "The Principle of Reason: The University in the Eyes of Its Pupils." *Diacritics* 13, no. 3 (1983): 6–20.

———. "Psyche: Inventions of the Other." In *Reading de Man Reading*, edited by Lindsay Waters and Wlad Godzich, 25–65. Minneapolis: University of Minnesota Press, 1989.

———. *Specters of Marx: The State of the Debt, the Work of Mourning, and the New International*. New York and London: Routledge, 1994.

————. *Speech and Phenomena and Other Essays on Husserl's Theory of Signs.* Evanston, IL: Northwestern University Press, 1973.

————. *Writing and Difference.* London: Routledge, 1980.

Derrida, Jacques, and Bernard Stiegler. *Echographies of Television: Filmed Interviews.* Cambridge: Polity Press, 2002.

Dijkstra, Edsger W. "Go to Statement Considered Harmful." *Communications of the ACM* 11, no. 3 (1968): 146–48.

Doyle, Richard. *Wetwares: Experiments in Postvital Living.* Minneapolis: University of Minnesota Press, 2003.

DuGay, Paul, Stuart Hall, Linda Janes, Hugh Mackay, and Keith Negus. *Doing Cultural Studies: The Story of the Sony Walkman.* London: Sage/The Open University, 1997.

Eckel, Bruce. *Thinking in C++.* Englewood Cliffs, NJ: Prentice Hall, 1995.

Fischer, Charles. N., and Richard J. LeBlanc. *Crafting a Compiler.* Menlo Park, CA: Benjamin/Cummings, 1988.

Fuller, Matthew. *Behind the Blip: Essays on the Culture of Software.* New York: Autonomedia, 2003.

————, ed. *Software Studies. A Lexicon*, Cambridge, MA and London: MIT Press, 2008.

Fuller, Matthew, and Andrew Goffey, eds. *Evil Media.* Cambridge, MA and London: MIT Press, 2012.

Fuller, Matthew, and Tony D. Sampson, eds. *The Spam Book: On Viruses, Porn, and Other Anomalies from the Dark Side of Digital Culture.* Cresskill: Hampton Press, 2009.

Galler, Bernard A. "Thoughts on Software Engineering." *Proceedings of the 11th International Conference on Software Engineering* 11, no. 2 (1989): 97.

Galloway, Alexander. *Protocol: How Control Exists after Decentralization.* Cambridge, MA: MIT Press, 2004.

Gell, Alfred. *Art and Agency*, Oxford: Clarendon Press, 1998.

————. "The Technology of Enchantment and the Enchantment of Technology." In *Anthropology, Art, and Aesthetics*, edited by Jeremy Coote and Antony Shelton, 40–63. Oxford: Clarendon Press, 1992.

Gille, Bertrand. *History of Techniques.* New York: Gordon, 1986.

Glass, Robert L. "Glass." *System Development*, January 1988.

————. *In the Beginning: Recollections of Software Pioneers.* Hoboken, NJ: John Wiley, 2003.

Gordon, Robert. M. "Review of *The Management of Computer Programming Projects*, by Charles P. Lecht." *Datamation* 14, no. 4 (1968): 11.

Gries, David. "My Thoughts on Software Engineering in the Late 1960s." *Proceedings of the 11th International Conference on Software Engineering* 11, no.2 (1989): 98.

Habermas, Jürgen. *The Future of Human Nature.* Cambridge: Polity Press, 2003.

————. *The Postnational Constellation.* Cambridge: Polity Press, 2001.

————. *The Theory of Communicative Action.* Cambridge: Polity Press, 1991.

Halberstam, Judith. *The Queer Art of Failure.* Durham and London: Duke University Press, 2011.

Hall, Gary. *Culture in Bits: The Monstrous Future of Theory.* London and New York: Continuum, 2002.

————. "The Digital Humanities beyond Computing: A Postscript." *Culture Machine* 12 (2011). www.culturemachine.net/index.php/cm/article/download/441/471.

————. *Digitize This Book! The Politics of New Media*, or *Why We Need Open Access Now.* Minneapolis: University of Minnesota Press, 2008.

———. "IT, Again: How to Build an Ethical Virtual Institution." In *Experimenting: Essays with Samuel Weber*, edited by Simon Morgan Wortham and Gary Hall, 116–40. New York: Fordham University Press, 2007.

Hall, Gary, and Clare Birchall, eds. *New Cultural Studies: Adventures in Theory*. Edinburgh: Edinburgh University Press, 2006.

Hall, Stuart. "Cultural Studies and Its Theoretical Legacies." In *Cultural Studies*, edited by Lawrence Grossberg, Cary Nelson, and Paula Treichler, 277–94. New York and London: Routledge, 1992.

———, ed. *Representation: Cultural Representations and Signifying Practices*. London: Sage/The Open University, 1997.

Hansen, Mark B. N. *New Philosophy for New Media*. Cambridge, MA: MIT Press, 2004.

———. "'Realtime Synthesis' and the Différance of the Body: Technocultural Studies in the Wake of Deconstruction." *Culture Machine* 5 (2003). http://www.culturemachine.net/index.php/cm/article/view/9/8.

Hansen, Mark B. N., and W. J. T. Mitchell, eds. *Critical Terms for Media Studies*. Chicago and London: University of Chicago Press, 2010.

Harel, David. "Biting the Silver Bullet: Toward a Brighter Future for System Development." *IEEE Computer* 25, no.1 (1992): 8–20.

Harrison, Harry, and Marvin Minsky. *The Turing Option*. London: ROC, 1992.

Hayles, Katherine K. *How We Became Posthuman: Virtual Bodies in Cybernetics, Literature and Informatics*. Chicago: University of Chicago Press, 1999.

———. *Writing Machines*. Cambridge, MA: MIT Press, 2002.

———. *My Mother Was a Computer: Digital Subjects and Literary Texts*. Chicago: University of Chicago Press, 2005.

———. *How We Think: Digital Media and Contemporary Technogenesis*. Chicago: University of Chicago Press, 2012.

Hayles, Katherine N., and Jessica Pressman, eds. *Comparative Textual Media: Transforming the Humanities in the Postprint Era*. Minneapolis and London: University of Minnesota Press, 2013.

Heidegger, Martin. *The Question Concerning Technology and Other Essays*. New York: Harper and Row, 1977.

Hicks, Harry T. "Modular Programming in COBOL." *Datamation* 14, no. 5 (1968): 50–59.

Hood, Webster J. "The Aristotelian Versus the Heideggerian Approach to the Problem of Technology." In *Philosophy and Technology: Readings in the Philosophical Problems of Technology*, edited by Carl Mitcham and Rober Mackey, 347–63. London: The Free Press, 1983.

Hopcroft, John. E., Rajeev M. Motwani, and Jeffrey. D. Ullman. *Introduction to Automata Theory, Languages, and Computation*. Reading, MA: Addison-Wesley, 2001.

Hopcroft, John E., and Jeffrey D. Ullman. *Formal Languages and Their Relation to Automata*. Reading, MA: Addison-Wesley, 1969.

Hopkins, Martin. "SABRE PL/I." *Datamation* 14, no.12 (1968): 35–38.

Huhtamo, Erkkiand, and Jussi Parikka, eds. *Media Archaeology: Approaches, Applications, and Implications*. Berkeley: University of California Press, 2011.

Humphrey, Watts S. *Managing the Software Process*. Harlow: Addison-Wesley, 1989a.

———. "The Software Engineering Process: Definition and Scope." In *Representing and Enacting the Software Process: Proceedings of the 4th International Software Process Workshop, 1989*, 82–83. ACM Press, 1989b.

Husserl, Edmund. *The Crisis of European Sciences and Transcendental Phenomenology. An Introduction to Phenomenological Philosophy.* Evanston, IL: Northwestern University Press, 1970.

Jones , Capers. *Estimating Software Costs.* New York: McGraw-Hill, 2007.

Kaaranen, Heikki, Ari Ahtiainen, Lauri Laitinen, Siamak Naghian, and Valtteri Niemi. *UMTS Networks: Architecture, Mobility and Services.* Chichester: John Wiley, 2005.

Kember, Sarah, and Joanna Zylinska. *Life after New Media: Mediation as a Vital Process.* Cambridge, MA and London: MIT Press, 2012.

Kirschenbaum, Matthew G. "Materiality and Matter and Stuff: What Electronic Texts Are Made Of." *Electronic Book Review* 12 (2002). http://www.altx.com/ebr/riposte/rip12/rip12kir.htm.

———. "Virtuality and VRML: Software Studies after Manovich." Electronic Book Review, 2003. http://www.electronicbookreview.com/thread/technocapitalism/morememory.

Kitchin, Rob, and Martin Dodge. *Code/Space: Software and Everyday Life.* Cambridge, MA and London: MIT Press, 2011.

Kittler, Friedrich. A. *Essays: Literature, Media, Information Systems.* Amsterdam: G+B Arts, 1997.

———. *Gramophone, Film, Typewriter.* Stanford, CA: Stanford University Press, 1999.

———. "There Is No Software." *CTheory*, October 1995. http://www.ctheory.net/articles.aspx?id=74.

Korhonen, Juha. *Introduction to 4G Mobile Communications.* Boston: Artech House, 2014.

Latour, Bruno. *We Have Never Been Modern.* Cambridge, MA: Harvard University Press, 1993.

Leroi-Gourhan, André. *Gesture and Speech,* Cambridge, MA: MIT Press, 1993.

Licklider, Joseph C. R. "Underestimates and Overexpectations." In *ABM: An Evaluation of the Decision to Deploy an Anti-Ballistic Missile System,* edited by Abram Chayes and Jerome B. Wiesner, 118–29. New York: Signet, 1969.

Lister, Martin, Jon Dovey, Seth Giddings, Iain Grant, and Kelly Kieran. *New Media: A Critical Introduction.* London and New York: Routledge, 2003.

Mackenzie, Adrian. *Cutting Code: Software and Sociality.* New York and Oxford: Peter Lang, 2006.

Mahoney, Michael. S. "Finding a History for Software Engineering." *Annals of the History of Computing* 26, no. 1 (2004): 8–19.

Manovich, Lev. (2001) *The Language of New Media.* Cambridge, MA and London: MIT Press.

———. *Software Takes Command: Extending the Language of New Media.* New York and London: Bloomsbury, 2013.

Marino, Mark. "Disrupting Heteronormative Codes: When Cylons in Slash Goggles Ogle AnnaKournikova." In *Digital Arts and Culture Proceedings.* University of California Irvine, 2009. http://escholarship.org/uc/item/09q9m0kn.

———. "Reading Exquisite Code: Critical Code Studies of Literature." In *Comparative Textual Media: Transforming the Humanities in the Postprint Era,* edited by Katherine N. Hayles and Jessica Pressman, 283–309. Minneapolis and London: University of Minnesota Press, 2013.

Marvin, Carolyn. *When Old Technologies Were New: Thinking about Electric Communication in the Late Nineteenth Century.* New York and Oxford: Oxford University Press, 1988.

Mateas, Michael. "Expressive AI: A Hybrid Art and Science Practice." *Leonardo: Journal of the International Society for Arts, Sciences and Technology* 34, no. 2 (2001): 147–53.

Meyrowitz, Joshua. *No Sense of Place: The Impact of Electronic Media on Social Behavior.* Oxford and New York: Oxford University Press, 1985.

Mills, Harlan. "Chief Programmer Teams, Principles, and Procedures." *IBM Federal Systems Division Report FSC 71-5108.* Gaithersburg, MD, 1971.

Montfort, Nick, Patsy Baudoin, John Bell, Ian Bogost, Jeremy Douglass, Mark C. Marino, Michael Mateas, Casey Reas, Mark Sample, and Noah Vawter. *10 PRINT CHR$(205.5+RND(1)); : GOTO 10.* Berkeley: MIT Press, 2013.

Montfort, Nick, and Ian Bogost. *Racing the Beam: The Atari Video Computer System.* Cambridge, MA and London: MIT Press, 2009.

Morowitz, Harold J. *The Emergence of Everything: How the World Became Complex.* Oxford and New York: Oxford University Press, 2002.

Morrison, Philip, and Emily Morrison, eds. *Charles Babbage and His Calculating Engines: Selected Writings by Charles Babbage and Others.* New York: Dover, 1961.

Naur, Peter, and Brian Randell, eds. *Software Engineering: Report on a Conference Sponsored by the NATO Science Committee, Garmisch, Germany, 7th to 11th October 1968.* Brussels (Belgium): NATO Scientific Affairs Division, 1969.

Norman, Donald A. *The Design of Everyday Things.* New York: Basic Books, 1988.

Ong, Walter. J. *Orality and Literacy: The Technologising of the Word.* London and New York: Routledge, 1982.

———. *Rhetoric, Romance and Technology: Studies in the Interaction of Expression and Culture.* Ithaca, NY: Cornell University Press, 1971.

Opler, Ascher. "Fourth-Generation Software." *Datamation* 13, no. 1 (1967): 22–24.

Parikka, Jussi. "New Materialism as Media Theory: Medianatures and Dirty Matter." *Communication and Critical/Cultural Studies* 9, no. 1 (2012): 95–100.

Parnas, David L. "A Technique for Software Module Specification with Examples." *ACM Communications* 15, no. 5 (1972): 330–36.

Parry, Richard. The Stanford Encyclopedia of Philosophy, s.v. "*Episteme* and *Techne*." 2003.http://plato.stanford.edu/archives/sum2003/entries/episteme-techne/.

Pinney, Christopher, and Nicholas Thomas, eds. *Beyond Aesthetics: Art and the Technologies of Enchantment.* Oxford and New York: Berg, 2001.

Plant, Sadie. "The Future Looms: Weaving Women and Cybernetics." In *Cyberspace/Cyberbodies/Cyberpunk: Cultures of Technological Embodiment,* edited by Mike Featherstone and Roger Burrows, 45–64. London: Sage, 1995.

———. *Zeros and Ones: Digital Women and the New Technoculture.* London: Fourth Estate, 1998.

Plato. *The Collected Dialogues.* Princeton, NJ: Princeton University Press, 1989.

———. *The Republic.* Cambridge: Cambridge University Press, 2000.

Poster, Mark. "High-Tech Frankenstein, or Heidegger Meets Stelarc." In *The Cyborg Experiments: the Extensions of the Body in the Media Age,* edited by Joanna Zylinska, 15–32. London and New York: Continuum, 2002.

Randell, Brian. "Memories of the NATO Software Engineering Conferences." *IEEE Annals of the History of Computing* 20, no.1 (1988): 51–54.

———. "Software Engineering in 1968." *Proceedings of the IEEE 4th International Conference on Software Engineering,* 1–10. Munich, Germany, 1979.

Raymond, Eric S. "The Cathedral and the Bazaar." 2001a. http://www.unterstein.net/su/docs/CathBaz.pdf.

———. *The Cathedral and the Bazaar: Musings on Linux and Open Source by an Accidental Revolutionary*. Cambridge, MA: O'Reilly, 2001b.

Reynolds, Richard. *Super Heroes: A Modern Mythology*. Jackson: University Press of Mississippi, 1994.

Ross, Douglas T. "The Nato Conferences from the Perspective of an Active Software Engineer." *Proceedings of the 11th International Conference on Software Engineering* 11, no. 2 (1989): 101–2.

Salomaa, Arto. *Formal Languages*. London: Academic Press, 1973.

de Saussure, Ferdinand. *Course in General Linguistics*. Peru, IL: Open Course Publishing, 1988.

Sayers, Dorothy L. *The Mind of the Maker*. London: Mowbray, 1994.

Sebesta, Robert W. *Concepts of Programming Languages*. London and New York: Pearson and Addison-Wesley, 2008.

Sedgwick, Eve Kosofsky. *Touching Feeling: Affect, Pedagogy, Performativity*. Durham, NC and London: Duke University Press, 2003.

Shaw, Mary. "Prospects for an Engineering Discipline of Software." *IEEE Software* 7, no. 6 (1990): 15–24.

———. "Remembrances of a Graduate Student." *Annals of the History of Computing, Anecdotes Department* 11, no. 2 (1989): 141–43.

Simondon, Gilbert. *Du mode d'existence des objects techniques*. Paris: Aubier, 2001.

———. *L'Individuation psychique et collective*. Paris: Aubier, 1989.

Sommerville, Ian. *Software Engineering*. Boston: Addison-Wesley, 2011.

Stiegler, Bernard. "Memory." In *Critical Terms for Media Studies*, edited by Mark B. N. Hansen and W. J. T. Mitchell, 64–87. Chicago and London: University of Chicago Press, 2010.

———. *Technics and Time, 1: The Fault of Epimetheus*. Translated by Richard Beardsworth and George Collins. Stanford, CA: Stanford University Press, 1998.

———. *Technics and Time, 2: Disorientation*. Translated by Stephen Barker. Stanford, CA: Stanford University Press, 2009.

———. *Technics and Time, 3: Cinematic Time and the Question of Malaise*. Translated by Stephen Barker. Stanford, CA: Stanford University Press, 2011.

———. "Technics of Decision: An Interview." *Angelaki* 8, no. 2 (2003): 151–68.

Tanenbaum, Andrew. S. *Structured Computer Organization*. Englewood Cliffs, NJ: Prentice-Hall, 1999.

Turner, John A. "Famous Fortran Errors." http://www.rchrd.com/Misc-Texts/Famous_Fortran_Errors.

Ullman, Ellen. *Close to the Machine: Technophilia and Its Discontents*. San Francisco: City Lights Books, 1997.

Virilio, Paul. *Open Sky*. London: Verso, 1997.

———. *The Vision Machine*. Bloomington: Indiana University Press, 1994.

Webb, Fred. "Famous Fortran Errors."http://www.rchrd.com/Misc-Texts/Famous_Fortran_Errors.

Williams, Raymond. *The Long Revolution*. Harmondsworth: Penguin, 1961.

Wirth, Niklaus. *Algorithms + Data Structures = Programs*. Englewood Cliffs, NJ: Prentice Hall, 1976.

Wolfram, Stephen. *A New Kind of Science*. New York: Wolfram Media, 2002.

Index

185